Geospatial Technology for Water Resource Applications

Geospatial Technology for Water Resource Applications

Editors

Prashant K. Srivastava

Hydrological Sciences (Code 617)
NASA GSFC
Greenbelt, Maryland
USA
and
Institute of Environment & Sustainable Development
Banaras Hindu University (BHU)
Varanasi
UP
India

Prem Chandra Pandey

Department of Geography
University of Leicester
University Road
Leicester
UK

Pavan Kumar

Department of Remote Sensing, Bhu Mandir
Banasthali University
Tonk, Rajasthan
India

Akhilesh Singh Raghubanshi

Institute of Environment & Sustainable Development
Banaras Hindu University (BHU)
Varanasi
UP
India

Dawei Han

Dept. of Civil Engineering
University of Bristol
Bristol
UK

CRC Press
Taylor & Francis Group
Boca Raton   London   New York

CRC Press is an imprint of the
Taylor & Francis Group, an **informa** business

A SCIENCE PUBLISHERS BOOK

Cover Acknowledgement

Satellite image in the background: European Space Agency

Classified satellite image (second image in the right panel): Disaster Prevention System Laboratory, Yamaguchi University, Japan.

CRC Press
Taylor & Francis Group
6000 Broken Sound Parkway NW, Suite 300
Boca Raton, FL 33487-2742

© 2017 by Taylor & Francis Group, LLC
CRC Press is an imprint of Taylor & Francis Group, an Informa business

No claim to original U.S. Government works

Printed on acid-free paper
Version Date: 20160510

International Standard Book Number-13: 978-1-4987-1968-1 (Hardback)

Library of Congress Cataloging-in-Publication Data

Names: Srivastava, Prashant K., editor.
Title: Geospatial technology for water resource applications / editors,
Prashant K. Srivastava, Prem Chandra Pandey, and Pavan Kumar.
Description: Boca Raton : Taylor & Francis, 2017. | Includes bibliographical
references and index.
Identifiers: LCCN 2016008973 | ISBN 9781498719681 (hardcover : alk. paper)
Subjects: LCSH: Water-supply--Remote sensing. | Water-supply--Geographic
information systems. | Water-supply--Management.
Classification: LCC GB1001.72.R42 G46 2017 | DDC 333.9100285--dc23
LC record available at https://lccn.loc.gov/2016008973

Visit the Taylor & Francis Web site at
http://www.taylorandfrancis.com

and the CRC Press Web site at
http://www.crcpress.com

Preface

◇◇

Information on our planet's water resources is essential for both human society and ecosystems. The world population is increasing at a rapid pace and the demand for water will continue to rise in future. However, the water resources are limited and therefore there is a growing need to monitor this resource and accumulate information in the technical literature domain that could assist the stakeholders towards development of effective water management strategies and infrastructures timely.

In recent decades geospatial techniques have gained considerable interest among the earth and hydrological science communities for solving and understanding various complex problems and approaches towards water resource development. This book is motivated by the desire to solve the problem of increasing scarcity of water resources in a cost effective and timely way. After the development of sophisticated geospatial technology and the launch of several Earth Observation (EO) satellites, it is now possible to monitor water resources regularly, accurately and in real time.

Overtime, geospatial techniques for water resources are becoming increasingly important in a much wider range of scientific and engineering disciplines. Looking at the importance of the subject, many departments across the world have revised their curricula and now have geo-informatics as a core subject along with the others like water resources or hydrology. Most introductory courses on geo-informatics are focused mostly on theory, so we have covered the practical applications of the geospatial techniques in this book, laid on a sound background of hydrology, and geospatial and computational intelligence techniques. Therefore, the main aim of this book is to provide an understanding of GIS, remote sensing and hybrid approaches to students and researchers in order to provide a solid foundation for further studies. The second goal is to provide readers with an insight into advanced courses such as geo-informatics and the conceptual tools that can be used in this field. Finally, a more pervasive goal of the book is to expose all students not only to advanced geospatial concepts but also to the intellectually rich foundations of the field.

After reading the literatures, we have found that it is useful to have a coherent architecture to motivate readers on how geospatial system works but have also found that almost all sets of courses revolve around separate architecture, which is not useful for providing a holistic framework and integration of the aspects. Foundations of geo-informatics cover subjects that are often split between sophisticated mathematical backgrounds and actual applications. Synergy is required between the two to select the mathematical foundations with an eye toward the demand of the users.

In order to simplify geospatial technology for most of the students and researchers, this book focuses on three working methodologies *viz* theory, abstraction and design as they are fundamental to all research programs. This book identifies the key recurring concepts which are fundamental to geospatial technology for water resources, especially: conceptual and formal models, efficiency and levels of abstraction. Following the working methodologies, processes and concepts, the primary aim of this book is to advance the scientific understanding, development and application of geospatial technologies to address a variety of issues related to water resource development. By linking geospatial techniques with new satellite missions for earth and environmental science oriented problems, this book will promote the synergistic and multidisciplinary activities among scientists and users working in the field of hydrological sciences. Many key topics are covered in the book from utilization of GIS, satellite based information to hybrid and artificial intelligence techniques for water resources.

This book has put together a collection of the recent developments and rigorous applications of the geospatial techniques for water resources. It will serve as the first handbook encompassing a spectrum of interests in water resources on a geospatial platform. We believe that the book will be read by the people with a common interest in geospatial techniques, remote sensing, sustainable water resource development, applications and other diverse backgrounds within earth and environmental and hydrological sciences field. This book would be beneficial for academicians, scientists, environmentalists, meteorologists, environmental consultants and computing experts working in the area of water resources.

This book is the result of extensive and valuable contributions from interdisciplinary experts from all over the world in the field of remote sensing, geospatial technologies, coastal science, ecology, environmental science, natural resources management, geography and hydrology representing academic, governmental and business sectors. The Editors are grateful to all the contributing authors and reviewers for their time, talent and energies.

About the Cover

Sophisticated Soil Moisture and Ocean Salinity satellite (Photograph provided by European Space Agency) is shown on the cover providing the state of the art soil moisture information from space. Other photos on the cover show the geospatial information contained by various layers and the land cover for water resources applications.

Maryland, USA   *Prashant K. Srivastava*
Leicester, UK   *Prem Chandra Pandey*
Rajasthan, India   *Pavan Kumar*
Varanasi, India   *Akhilesh Singh Raghubanshi*
Bristol, UK   *Dawei Han*

Contents

Section V   Challenges in Geospatial Technology For
                Water Resources Development

List of Contributors

◇◇

Muhammad Zulkarnain Abd Rahman

TropicalMap, Faculty of Geoinformation & Real Estate, Universiti Teknologi Malaysia, 81310 UTM, Johor Bahru, Johor, Malaysia

Muhammad Ali

Department of Civil Engineering, University of Engineering and Technology, Taxila-47050, Pakistan

Putu Aryastana

Civil Engineering Department, Warmadewa University, Jl Terompong No 24 Tanjung Bungkak, Denpasar, Indonesia

Muhammad Saadoon Basraa

Department of Civil Engineering, University of Engineering and Technology, Taxila-47050, Pakistan

Madan Raj Bhatta

Water Resources Management Program (WARM-P)/HELVETAS, Swiss Intercooperation Nepal, Dhobi Ghat, Lalitpur GPO Box 688, Surkhet, Nepal

Ngai-Weng Chan

School of Humanities, Universiti Sains Malaysia, 11800 USM Penang, Malaysia

Qiang Dai

Key Laboratory of VGE of Ministry of Education, Nanjing Normal University, Nanjing-210046, China

Sandeep Kumar Gautam

School of Environmental Sciences, Jawaharlal Nehru University, New Delhi–110067, India

Dileep Kumar Gupta

Department of Physics, Indian Institute of Technology (B.H.U.), Varanasi-221005, India

Manika Gupta
Hydrological Sciences, NASA Goddard Space Flight Center, Greenbelt-20771, Maryland, USA

Dawei Han
Water and Environmental Management Research Centre, Department of Civil Engineering, University of Bristol, Bristol BS8 1TR, UK

Ab Latif Ibrahim
Geoscience and Digital Earth Centre, Research Institute of Sustainability and Environment, Universiti Teknologi Malaysia, 81310 UTM, Johor Bahru, Johor, Malaysia

Moslem Imani
Department of Geomatics, National Cheng Kung University, No. 1, University Road, Tainan City, 701, Taiwan

Tanvir Islam
NASA Jet Propulsion Laboratory, Pasadena, CA-91109, USA; and California Institute of Technology, Pasadena, CA-91125, USA

Swati Katiyar
Department of Remote Sensing, Banasthali University, Newai, Tonk-304022, Rajasthan, India

Anand N. Khobragade
Resource Scientist, Maharashtra Remote Sensing Applications Center, Nagpur-440010, India

Pavan Kumar
Department of Remote Sensing, Bhu Mandir, Banasthali University, Newai, Tonk, Rajasthan 304022, India

Rajesh Kumar
Centre for the Study of Regional Development, School of Social Sciences, Jawaharlal Nehru University, New Delhi-110067, India

S. Prasanna Kumar
Department of Physical Oceanography, National Institute of Oceanography, Dona Paula, North Goa 403004, India

Sanjay Kumar
Rural Development Officer, Govt. of Bihar, Patna-800015, Bihar, India

Chung-Yen Kuo

Department of Geomatics, National Cheng Kung University, No. 1, University Road, Tainan City, 701, Taiwan

Jia Liu

State Key Laboratory of Simulation and Regulation of Water Cycle in River Basin, China Institute of Water Resources and Hydropower Research, 971#, Building A, No.1-A Fuxing Road, Beijing 100038, China

Panagiota Louka

Department of Natural Resources Development and Agricultural Engineering, Agricultural University of Athens, Iera Odos 75, Athina 11855, Greece

Vinay Mandal

Research Associate, PDFSR Modipuram, Meerut, 250110, Uttar Pradesh, India

Fusanori Miura

Graduate School of Environmental Science and Engineering, Yamaguchi University, Tokiwadai 2-16-1, Ube, Yamaguchi, Japan

Abdul Moiz

Department of Civil Engineering, University of Engineering and Technology, Taxila-47050, Pakistan

Usman Ali Naeem

Department of Civil Engineering, University of Engineering and Technology, Taxila-47050, Pakistan

Boini Narsimlu

Division of FM and PHT, ICAR–Indian Grassland and Fodder Research Institute, Jhansi-284003 (U.P.), India

M.S. Nathawat

School of Life Sciences, Indira Gandhi National Open University, New Delhi-110030, India

Mahesh Neupane

Department of Water Supply and Sewerage (DWSS), Government of Nepal, Kathmandu-32900, Nepal

Prem Chandra Pandey

Centre for Landscape and Climate Research, Department of Geography, University of Leicester, Leicester LE1 7RH, UK

Ioannis Papanikolaou

Department of Natural Resources Development and Agricultural Engineering, Agricultural University of Athens, Iera Odos 75, Athina 11855, Greece

George P. Petropoulos

Department of Geography & Earth Sciences, University of Aberystwyth, Aberystwyth SY23 3FL, Wales, UK

Rajendra Prasad

Department of Physics, Indian Institute of Technology (B.H.U.), Varanasi-221005, India

Akhilesh Singh Raghubanshi

Institute of Environment and Sustainable Development, Banaras Hindu University, Varanasi 221005, UP, India

Ragindra Man Rajbhandari

NEST Pvt. Ltd., Shankhamul, Kathmandu-44600, Nepal

Indal K. Ramteke

Scientific Associate, Maharashtra Remote Sensing Applications Center, Nagpur-440010, India

Preeti Rani

Department of Physical Oceanography, National Institute of Oceanography, Dona Paula, North Goa 403004, India

N. Ravishankar

PDFSR, Modipuram, Meerut, 250110, Uttar Pradesh, India

Miguel A. Rico-Ramirez

Water and Environmental Management Research Centre, Department of Civil Engineering, University of Bristol, Bristol-BS8 1TR, UK

A. Besse Rimba

Graduate School of Environmental Science and Engineering, Yamaguchi University, Tokiwadai 2-16-1, Ube, Yamaguchi, Japan

Martiwi Diah Setiawati

Graduate School of Science and Engineering, Department of Environmental Science and Engineering, Yamaguchi University, 2-16-1 Tokiwadai, Ube, 755-8611, Japan

Muhammad Ali Shamim

Department of Civil Engineering, Bursa Orhangazi University, Bursa-16310, Turkey

Rubika Shrestha

Water Resources Management Program (WARM-P)/HELVETAS, Swiss Intercooperation Nepal, Dhobi Ghat, Lalitpur GPO Box 688, Surkhet, Nepal

Abhay K. Singh

Central Institute of Mining and Fuel Research, Barwa Road, Dhanbad, Jharkhand-826001, India

Prafull Singh

Amity Institute of Geo-Informatics and Remote Sensing, Amity University-Sector 125, Noida-201303, India

Sudhir Kumar Singh

K. Banerjee Centre of Atmospheric and Ocean Studies, IIDS, Nehru Science Centre, University of Allahabad, Allahabad-211002 (U.P.), India

Prashant K. Srivastava

Hydrological Sciences (Code 617), NASA GSFC/JPL, Room No. G208, Building 33, Greenbelt, Maryland 20771, USA; Institute of Environment and Sustainable Development, Banaras Hindu University, Varanasi 221005, UP, India; and Earth System Science Interdisciplinary Center, University of Maryland, College Park-20742, Maryland, USA

Nikolaos Stathopoulos

Section of Geological Sciences, School of Mining Engineering, National Technical University of Athens, Athens-10682, Greece

Tze-Huey Tam

Geoscience and Digital Earth Centre, Research Institute of Sustainability and Environment, Universiti Teknologi Malaysia, 81310 UTM, Johor Bahru, Johor, Malaysia

Jay Krishna Thakur

Environment and Information Technology Centre–UIZ, Neue Grünstrasse 38, 10179, Berlin, Germany

Bikram Rana Tharu

Building Effective Water Governance in Asian Highlands (HELVETAS Swiss Intercooperation Nepal), Dhobi Ghat, Lalitpur GPO Box 688, Kathmandu, Nepal

Vandana Tomar

Haryana Institute of Public Administration (HIPA), 76, HIPA Complex, Sector-18, Gurgaon, Delhi-122001, India

Jayant K. Tripathi

School of Environmental Sciences, Jawaharlal Nehru University, New Delhi–110067, India

Wan Zurina Wan Jaafar

Department of Civil Engineering, Faculty of Engineering, University of Malaya, 50603 Kuala Lumpur, Malaysia

Iain Woodhouse

School of Geosciences, The University of Edinburgh, Edinburgh EH9 8XP, UK

Aradhana Yaduvanshi

Center of Excellence in Climatology, Birla Institute of Technology, Mesra-835215, Ranchi, India

Rey-Jer You

Department of Geomatics, National Cheng Kung University, No. 1, University Road, Tainan City, 701, Taiwan

Section I

General

1

CHAPTER

◇◇

Introduction to Geospatial Technology for Water Resources

Prem Chandra Pandey,[1] *Prashant K. Srivastava,*[2,3,4,]* *Pavan Kumar,*[5] *Akhilesh Singh Raghubanshi*[4] and *Dawei Han*[6]

ABSTRACT

Increasing demands on water resources to fulfill the growing population needs have led to a great pressure on the water resources. Water resources conservation and management needs exemplary information regarding the water bodies with respect to quality, quantity and the related driving factors responsible for deterioration and depletion of water. Traditional methods existing in literature are limited to the point locations and manually gathered input dataset for analysis of the water system. However, after the development of advance geospatial technologies, now it is possible to build the digital information that can support analysis and interpretation for a large area in short span of time. The chapter introduces the various geospatial technologies, which are playing a vital and inevitable role in the acquisition of information and development of research capabilities towards water resources. These technologies are required for determining a strategic plan for execution of desired results as applicable to different regions and objectives (for e.g. determination of water-river boundaries, water quality and quantity, soil moisture, flood plains, ocean temperature etc). This chapter provides different methods/applications to demonstrate the importance of traditional and advanced concepts of geospatial technology in water resources. Thus, overall goal

[1] Centre for Landscape and Climate Research, Department of Geography, University of Leicester, Leicester, UK.
[2] Hydrological Sciences, NASA Goddard Space Flight Center, Greenbelt, Maryland, USA.
[3] Earth System Science Interdisciplinary Center, University of Maryland, Maryland, USA.
[4] Institute of Environmental and Sustainable Development, Banaras Hindu University, Varanasi, India.
[5] Department of Remote Sensing, Banasthali University, Newai, Tonk, Rajasthan, India.
[6] Water and Environmental Management Research Centre, Department of Civil Engineering, University of Bristol, Bristol, UK.
* Corresponding author: prashant.just@gmail.com

of this chapter is to provide a summary of different research work carried out in various fields of water resources with demonstrated results and findings that could be able to use in decision making, developing policy and planning at root level. This chapter also provides future challenges in water resources and geospatial technology.

KEYWORDS: Water Resources, Geospatial Technology, Traditional methods, Advance methods, Hybrid Technology, Challenges.

○ INTRODUCTION

Water is an essential component of natural resources available in many forms such as groundwater, river, springs, lakes, glaciers etc. The importance of water comes from its ability to keep us alive, carry nutrients to crops and plants, dilute wastes and toxic-pollutants and maintain the hydrological cycle. Therefore, water resources management is a significant issue for us today in order to reduce water scarcity for future generations. Nowadays, because of substantial demographic and economic changes there are high fluctuations in the hydrological regime that cause depletion or contamination of water resources. As a result, this precious resource is under pressure and needs conservation management as well as protection.

The vital issue of understanding water resources through geo-informatics is addressed in this book through traditional, advanced, hybrid and artificial intelligence techniques. Water resources management is mainly used to understand and monitor water resources, while geospatial techniques use RS, GIS and GPS applications to facilitate this practice in a timely fashion. Therefore, the aim of this book is to present several ideas towards monitoring of water resources. Apart from its role as a life supporter, water also causes harmful hazards in the form of flood and drought. This book contains several techniques for monitoring and protection of water resources to hazard assessment such as frost, floods and droughts. It also outlines future challenges that need attention for us to cope up with the management of this precious natural resource.

This book is divided into four sections to cover the water resources applications using Remote Sensing, GIS, Hybrid and Artificial intelligence techniques and also provide challenges which need to be addressed in the near future. Section Two provides the geographical information based application on water resources such as flood risk, nutrients for irrigation of agricultural crops, hydropower and watershed management. Chapter 2 provides the master plan for the integrated water resources management using geographical information system. The principle concept of 3R i.e. recharge, retention and reuse are presented in this chapter. This 3R concept with integrated GIS analysis can be used in calculating the current and future hardships for water in catchments. This idea helps in identifying the potential opportunities for critical watershed and hardship zones and implement watershed conservation activities for balancing ecosystem services for water and land use.

Chapter 3 elucidates the spatial integration of soil and water nutrients which are essential for crop growth and production. This chapter presents the sodium absorption ratio (SAR) with different nutrients including nitrogen, phosphorus and potassium for agricultural applications. The different water parameters (pH, organic carbon etc) are used in the spatial mapping of the spatial discrepancies of rice crop based on cropping systems productivity. Finally, this chapter provides the relational pattern in rice equivalent field and nitrate mapping. Chapter 4 focuses on assessment of hydropower potential using Shuttle Radar Topography Mission (SRTM) based Digital Elevation Model (DEM) along with observed discharge. Several locations of river hydropower plants are identified to obtain the results in GIS environment. Results from this study could be helpful in making precise decisions in favour of a sustainable hydropower development and finally alleviate the energy crisis being faced today by the Pakistan. The study also indicates the usefulness of using GIS in carrying out spatial interpolation and consequently, determination of hydropower potential. This chapter apprehends the energy demand due to an increasing population, which is putting a high pressure on the economic growth and proved the importance of renewable energy sources as a prerequisite for sustainable economic growth.

Chapter 5 introduces the flood risk assessment using different techniques and also provides the delineation and zonation of flood risk areas through geo-hydrological parameters. They generate flood risk maps using satellite LiDAR, hydrological data and two dimensional hydraulic models. The authors have presented different techniques to estimate the flood in the residential areas. This chapter has effectively produced the flood risk maps and suggested ideas to improve them by taking into account the epistemic uncertainties. The risk and hazard maps generated in this study will help in reducing socio-economic losses. In Chapter 6, authors have used multi-temporal multi-source images along with secondary data sets such as the inundated and flood atlas to identify flood prone areas. They have delineated three flood zones using the overlay of flood layers of different magnitude and geomorphic features in the GIS environment. Authors have also attempted the categorization of areas into alluvial islands and mid-channel bars, protected embankment, older floodplain with meander scars and scroll bars. The authors have shown the increasing nature of hamlet density in relation to future flood events in the lower Ghaghara river valley. Chapter 7 provides the geospatial approach for water resource management in a watershed using the merged ortho-product of LISS-IV and Cartosat-1D satellites. The authors build water resource development plans by considering the various criteria themes such as slope, soil and land use/land cover. The aim of the chapter is to provide a better use of natural resources by conserving both soil and water resources. The chapter basically deals with the developmental activities to provide soil and water conservation measures by reducing scarcity, erosions, agricultural development as well as to improve ground water.

The third section of the book mainly deals with the satellite based applications in water resources, where flood vulnerable areas, frost risk, precipitation monitoring for flood assessment, surface and ground water for irrigation purposes and sea surface water height anomaly are presented. In this section, Chapter 8 deals with the flood vulnerability prediction using satellite and Amedas data derived products in Japan. The case studies presented in this chapter include the Shiragawa watershed, Japan to predict the flood vulnerable regions. The isohyet map prepared using rain-gauge data is interpolated with kriging applications. Finally, all spatial data are overlaid to create the flood-vulnerability map by application of the GIS model. Chapter 9 deals with the estimation of precipitation for flood monitoring and its validation using the hourly GSMaP (Global Satellite Mapping Precipitation) satellite datasets with MVK (Moving Vector with Kalman Filter), NRT (Near Real Time) data and 27 rain gauges ground based reference station datasets (AMEDAS -Automated Meteorological Data Acquisition System). The study is carried out at spatial resolution 0.1° latitude × 0.1° longitude in Kumamoto Japan, often prone to flash flood, for the temporal duration of 2003 to 2012. The main theme of the chapter is to define the rainfall pattern causing flash flood using statistical analysis. Chapter 10 presents an interesting topic related to qualitative assessment of surface and ground water evaluation using remote sensing, irrigation indices and statistical techniques. The chapter revolves around the main theme of irrigation and drinking suitability of surface and groundwater. To meet the objective, average concentration of several water parameters is determined in the surface and the groundwater samples and presented using geospatial techniques. Chapter 11 discusses the techniques for assessment of spatial and temporal variation of sea surface height anomaly and its relationship with satellite derived chlorophyll-a pigment concentration. The authors derived the sea surface height anomaly (SSHA) data using satellite altimetry and chlorophyll-a concentration from ocean colour, which are subjected to Empirical Orthogonal Function (EOF) analysis to understand the spatio-temporal variability of these parameters in the context of mesoscale eddies. They suggest EOF as an efficient method of delineating a spatial and temporal signal from a long time series data over a large spatial domain. Chapter 12 emphasizes soil moisture deficit monitoring using the Soil Moisture and Ocean Salinity (SMOS) Satellite through rainfall-runoff model. Several approaches for estimation of soil moisture deficit are performed using SMOS satellite soil moisture through Generalized Linear Model with different families/link functions such as Gaussian/logit, Binomial/identity, Gamma/inverse and Poisson/log. The overall performances obtained from all the techniques indicate that the SMOS is promising for simulation of soil moisture deficit.

Section Four focuses on the artificial intelligence and hybrid models for water resource for flood regionalization, soil moisture monitoring, sea level prediction and spatio-temporal uncertainty model for rainfall predictions. Chapter 13 reveals the techniques to measure frost risk and suggests techniques to avoid frost risk in agriculture using MODIS data. The authors present the spatio-temporal distribution of frost conditions in Mediterranean environments. They present a model based on the main factors that include environmental factors such as land surface temperature and geomorphology governing the frost risk. Several topographical parameters such as altitude, slope, steepness, aspect, topographic curvature and extent of the area influenced by water bodies are required in the model to assess the frost risk. MODIS and ASTER polar orbiting sensors, supported also by ancillary ground observation data along with land use and vegetation classification (i.e. types and density) are the required input for the successful determining of the frost risk for the winter period of the four different selected years. Overall, the proposed methodology proves to be capable of detecting frost risk in Mediterranean environments in a time efficient and cost effective way, making it a potentially very useful tool for agricultural management and planning. The proposed model may be used as an important tool for frost mapping, a natural hazard that leads to severe vegetation damage and agricultural losses. Chapter 14 provides a statistical approach for catchment calibration data in flood regionalization. The chapter concluded that the quantity and quality of calibration data are two different entities that could greatly influence the developed hydrological model. The study has demonstrated that the standard deviation values between the best and poorest groups are distinctive and could be used in choosing appropriate calibration catchments. Chapter 15 predicts the Caspian Sea level fluctuations using artificial intelligence and satellite altimetry. This chapter presents different approaches for studying the sea level anomaly, fluctuation and analysis. Accurate prediction of sea level is important as it affects the natural processes occurring in the basin and influences the infrastructure built along coastlines. Several conventional linear regression methods such as routine Autoregressive Moving Average (ARMA) models, neural network methodologies and artificial intelligence approaches and techniques are included in the study. Based on the results, the authors support the Support Vector Machine as the best performance technique in predicting sea level. Chapter 16 familiarises readers with a novel method of spatio-temporal uncertainty model based on the distribution of gauge rainfall conditioned on radar rainfall (GR|RR). This fully formulated uncertainty model has statistically quantified the characteristics of radar rainfall errors and their spatial and temporal structure. Its spatial and temporal dependencies are simulated based on a Multivariate Distributed Ensemble Generator driven by the copula and autoregressive filter designed including the different wind conditions.

In Chapter 17 authors investigated the bistatic scattering coefficients for the estimation of soil moisture over rough surfaces using fuzzy logic and bistatic scatterometer data at X-band. Linear regression analysis is carried out between scattering coefficients and soil moisture to find the suitable incidence angle for the estimation of soil moisture at HH and VV co-polarization. The last and final Chapter 18 forms the last section of this book, which presented the challenges in geospatial technology for water resources development. This chapter also provide the importance and value of Remote Sensing and Geographical Information System (GIS) in water resources and pointed out the challenges in this field that should be addressed by researchers, policy makers and practitioners in order to view the technology in proper perspective. Several challenges such as monitoring soil, snow and vegetation; soil moisture estimation; monitoring evapotranspiration and energy fluxes; uncertainties in retrieval algorithms; bias associated with instruments; short time spans of satellite data etc are identified that needs to be addressed in near future.

Section II

Geographical Information System Based Approaches

2

CHAPTER

❖❖❖

GIS Supported Water Use Master Plan: A Planning Tool for Integrated Water Resources Management in Nepal

Mahesh Neupane,[1,*] *Madan Raj Bhatta,*[2] *Rubika Shrestha,*[2]
Jay Krishna Thakur,[3] *Ragindra Man Rajbhandari,*[4]
and *Bikram Rana Tharu*[5]

ABSTRACT

A Water Use Master Plan (WUMP) is a holistic approach for integrated water resources management. The application of a geographical information system (GIS) in WUMP offers a spatial representation of the water resources in various social, economic, hydrological and geographic settings to better integrate recharge, retention and reuse (3R) principles into the WUMP process. This integrated analysis from GIS can be used in calculating the current and future hardships considering the possible impacts of climate change in the catchment area. This helps in identifying the potential opportunities for critical watersheds and hardship zones and to implement watershed

[1] Department of Water Supply and Sewerage (DWSS), Government of Nepal, Kathmandu, Nepal.
[2] Water Resources Management Program (WARM-P)/HELVETAS, Swiss Intercooperation Nepal, Surkhet, Nepal.
[3] Environment and Information Technology Centre – UIZ, Neue Grünstrasse 38, 10179, Berlin, Germany.
[4] NEST Pvt. Ltd., Shankhamul, Kathmandu, Nepal.
[5] Building Effective Water Governance in Asian Highlands (HELVETAS Swiss Intercooperation Nepal) Kathmandu, Nepal.
* Corresponding author: maheshneu@gmail.com

conservation activities for balancing ecosystem services such as water and land use from anearly stage in the planning process. This will further help for designing sustainable water resources projects in future considering the climate resilience against drying of hill springs simultaneously. GIS is a very crucial supporting tool in decision making processes, especially when sustainability of water supply systems and services are the major concerns, either from the investment or environment conservation perspective. HELVETAS' Water Resource Management Program (WARM-P) has prepared more than 100 WUMPs since 2001 in Nepal. A GIS supported WUMP focused on 3R is the further innovation of WARM-P for an integrated water resources management approach in Nepal. This paper presents the pilot implementation of GIS supported, 3R focused WUMP in Padukasthan village development committee (VDC) of Dailekh district in 2013.

KEYWORDS: water use master plan (WUMP), GIS, integrated water resources management (IWRM), 3R, climate change.

○ INTRODUCTION

Integrated water resources management (IWRM) has become critical in water resource management as it is the only sustainable way of meeting water demands and preserving the basic ecosystem services of nature (Mukherjee et al. 2007; Thakur and Thakur 2011; Diwakar and Thakur 2012). There is a need to ensure access to safe drinking water in a sustainable way (Neupane et al. 2014). In 1992 at the United Nation's Conference for Environment and Development (UNCED) in Rio de Janeiro, the holistic water resource approach referred to as the Dublin-Rio principle was discussed. It highlighted that fresh water is finite, vulnerable and that it is essential to sustain life, economic development and the environment (Snellen and Schrevel 2004). Water development and management should be based on a participatory approach, involving users, planners and policy makers at all levels (ICIMOD, HELVETAS and WOCAT (2013) (Durham et al. 2002).

A WUMP is a holistic, participatory and inclusive planning process that takes an integrated approach to the management of water resources and demands at the local level (Bhatta and Bhatta 2011; Gupta and Srivastava 2009; Nizami et al. 2013). The WUMP identifies the total water budget from different sources available at the smallest geographical unit (ward level), prioritizes them and finds out their scope for addressing the water needs of the community. It is a transparent process which empowers marginalized groups to claim their rights to an equitable share of water within and between communities.

Water use master plans (WUMPs) and a participatory step-by-step approach, both backed up by a gender equality and social inclusion strategy, ensure appropriate and fair use of scarce water resources (HELVETAS 2009; Rautanen and White 2013). Key project entry points thus include decentralization, participation and empowerment (Rautanen et al. 2014). The local planning unit in Nepal which has the lowest administrative boundary is called a village development committee. This group takes the lead role and ownership during the planning process and implementation. The WUMP helps the local body to carry out annual and periodic planning of water projects. However, practical approaches and innovative solutions such as a 3R approach (recharge, retention and reuse) are needed for the effective management of water resources and adaption to climate change. Thus, the WARM-P project piloted a 3R based WUMP in Padukasthan VDC of Dailekh district in 2013 for counteracting the potential water stress to be incurred due to the possible impacts of climate change. 3R based WUMP is an approach adapted by incorporating the principles of recharge, retention and reuse (3R) in operational water use planning. It can be used for climate change adaptation and enhanced water resources management. It can form the basis of rural planning for natural resources management for sustainable livelihood (Dent et al. 2013) and the framework for implementation of a sustainable system for rural communities in developing countries (Neupane et al. 2014).

○ PILOTING OF 3R BASED WUMP IN NEPAL: IN SUPPORT OF GIS APPLICATIONS (CASE STUDY)

Nationwide survey of water supply and sanitation schemes revealed the poor functionality of drinking water and sanitation schemes. According to the survey, 25.4 percentage are functioning well, 36.1 percentage need minor repairs, 9.2 percentage need major repairs, 19.8 percentage need rehabilitation, 8.6 percentage need reconstruction and 0.9 percentage are non-refunctionable (NMIP/DWSS 2014). Both the growing demand of water for domestic and production uses as well as the water stress issues induced by climate change is emerging as a key challenge for the water sector. Water Resources Act, 1992 of Government of Nepal has set up priorities for water sources based upon their uses with the key priority on drinking water followed by irrigation, hydropower and others. The functionality study conducted by WARM-P in 2011 in 92 gravity water supply schemes of the same region revealed that water yield decreased in 24% (out of 224 in 92 schemes) of the water sources as compared to the yield at the time of construction of schemes. During discussions with community members, as a part of the planning phase in WUMP, a majority of them stated that the primary water sources which they are using for their drinking purposes are drying up and they are forced to look for other sources to meet their water supply demand. Many existing water supply systems are in danger of abandonment due to drying up of their sources or inadequate water at the intake point. The WUMP + 3R approach addresses these issues with a focus on participatory planning of water resources and integration of 3R solutions in the water resources development plan.

The integration of GIS application in a 3R based WUMP facilitates the hydrological study of the catchments and helps to identify the critical water resources and associated catchment areas (Thakur. et al. 2012). GIS layers give a clear spatial whereabouts of the existing water resources (Srivastava et al. 2012) combined with various physical environments like catchment, hydrological basins, drainage networks (Singh et al. 2013a), contours, slope, aspect, land-use, available water budget and settlement, which further help to classify zones based on the hardship and future possible water stress zones which may be affected by the impacts of climate change. Consequently, different 3R technologies and other watershed conservation activities can be concentrated for balancing ecosystem services in those identified zones and thus ultimately strengthening the yield from those sources. Thus, it proves to be very beneficial tool in decision making during the planning process, so that either the conservation activities can be better utilized or the project implementers have a better understanding of the future scenarios and critical areas of different catchments prior to planning of the projects. Thus, transforming environmental data to information helps build knowledge to decision support and ultimately to social impact (Avtar et al. 2012; Srivastava et al. 2012). Spatial technologies are crucial not only in mapping water resources, especially water channels in a watershed, but also in land cover/land use change detections for general management of such resources (Nathanail 2013; Kaburi and Odero 2014).

○ STUDY AREA

The Padukastan VDC is located in Dailekh district of Bheri zone of Mid-Western Development Region of Nepal. The VDC is situated at 7 hr. walking distance from the district headquarter and is located in a remote hilly setting. The entire VDC is located on an extended hill at an average altitude of 1220 m from sea level. The village lies between longitude 81°32′ E to 81°37′ E and latitude 28°52′ N to 28°54′ N. The VDC can be divided into three sub-basins based upon the water flow and catchment peripheries and into nine Wards (smallest administrative boundaries), 61 clusters and 1007 households (HH). A cluster has been defined based upon the group of community who drink water from same drinking water source. Padukastan receives average annual rainfall of 1790 mm from June to September and the water catchment comprises of 129 hill springs and seven different streams. The existing water sources have been used for two different purposes; mainly as drinking water and for irrigation as well. A total 31.5 hectare of land is irrigated featuring mostly terrace farming.

The project site case study area of Padukasthan is shown in Fig. 1. The sources existing on the ridge side of a south facing aspect of a hill on steep slope may not be a better choice as compared to the sources existing on the valley side of north facing aspect on a mild slope of the same hill having higher discharge. Mild slopes favor the recharge or infiltration of flowing surface water and the northern aspect of a hill in northern hemisphere is a hill-shade zone and has high moisture and low evaporation loses. This is shown in Fig. 2 where the water sources are mostly located in the northern aspect in the hill-shade zone. All of these features can be clearly depicted through GIS layers in a single map (Patel et al. 2012; Vyas and Pandya 2013) which makes it easier to identify the potential 3R interventions in critical watersheds. Subsequently, a good selection of the 3R technologies based upon their salient features and appropriateness with respect to topography and geology can be suggested.

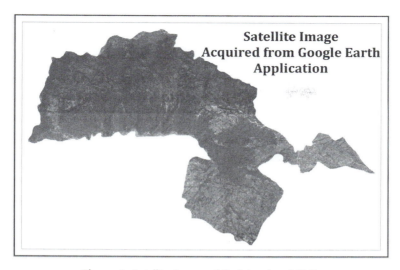

Figure 1 Satellite image of Padukasthan VDC.

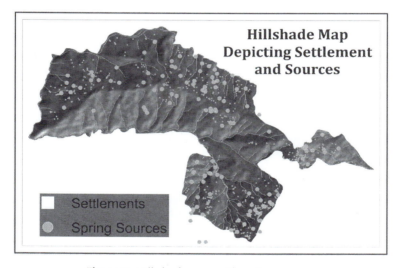

Figure 2 Hill shade map with water sources.

○ DESIGN AND METHODOLOGY

The main focus of the 3R based WUMP was to pilot the effectiveness of 3R measures in the existing WUMP which were identified through GIS analysis. 3R based WUMP involves analyzing every water source with respect to its location and topographic attributes such as aspect, slope and land cover through the use of GIS. The methodology for preparation of 3R based WUMP comprises five phases: Preparatory, Capacity Development, Assessment, Planning and Implementation as shown in Fig. 3 and in accordance with Recharge, Retention and Reuse based

Water Use Master Plan of Padukasthan VDC (Padukastan VDC, 2013). These five phases are further broken down into 17 steps and various sub-steps. In addition, GIS application and assessment of 3R measures were synchronized in the Assessment phase.

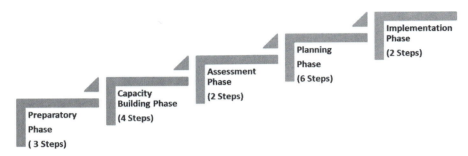

Figure 3 Phases and steps of WUMP process.

The entire 3R based WUMP is the outcome of a series of activities that were carried out in a well-defined stepwise process. Such activities included desktop works; capacity building; institutional development; field level activities; planning and prioritization of the schemes at VDC level; endorsement and dissemination; and marketing as major activities. Desktop works included review of available secondary data, analysis of the available data, demand-supply study and report preparation. Field level activities included social and resources mapping which involved primary data collection on demography, health & sanitation, education, water sources, existing water facilities and service level (fetching time, quantity, quality, reliability and continuity), inventory assessment of existing and on-going schemes, needs identification and prioritization of schemes at the sub-committee level. Special formats for data collection and guidelines for each step were developed prior to study.

The data on available water sources, present use of such sources and potential future uses with the needs of local people were identified at the ward level which is the lowest administrative boundary within the VDC. The collected data was analyzed and a draft report was prepared. The draft report was presented and discussed in the VDC level planning workshop to verify the data as well as prioritize for the planning of water schemes that came from the ward level survey.

Various participatory tools were applied for the delineation of the sub watershed, resources available and needs identification of local people. The prioritization for the use of net available water resources were completed as per the Water Resources Act, 1992 of Nepal which entitles drinking water as the first priority, followed by irrigation, hydropower and others. The data used in the planning was with participation of local communities. The field survey data was collected in the assessment phase and GIS was applied to record and analyze the information collected during this phase. The primary data was collected in the third phase whilst carrying out the two assessments listed below:

- Socio economic assessment and needs identification.
- Technical assessment.

TABLE 1 Technical assessment.

S. No.	Sub-step	Methodology	Participants	Data collected	Responsibility
1	Secondary data review	Data analysis/Desk study	WUMP facilitator	Related existing available data, like HHs, population, hydrology, different water schemes and its coverage etc.	Service provider (WUMP facilitator of technical background) with support of ward citizen forum
2	Water resources inventory and discharge measurement	Field mapping and measurements	Key informants from community, WUMP facilitator	Details of water sources (yield, elevation, access etc.)	
3	Inventory of water infrastructures	Field mapping and measurements	Key informants from community, WUMP facilitator	Existing water facility or drinking water systems (DWS), irrigation, water energy: Hydropower etc.) and its coverage	
4	Pre-feasibility assessment of potential schemes	Field observations and measurements	Key informants from community and WUMP facilitator	Proposed water schemes (DWS, irrigation, water energy: Hydropower, multiple use schemes like water supply and irrigation, irrigation and hydropower, water meal etc.)	

During the assessment phase, socio economic assessment was carried out using various participatory rural appraisal tools such social resource mapping, transect walk and participatory need identification. The assessment was carried out for each of the nine wards. For the situational analysis of the existing water supply condition, the data was collected for each cluster.

The socio economic assessment was followed by the technical assessment in each ward. The technical assessment was based on the physical inspection of available resources and different aspects within the catchment area which included: inventory of all water resources and their discharge measurement with GPS data; inventory of existing water related infrastructure; pre-feasibility of the potential water schemes; water and sanitation profile and service level and specific 3R measures. The-sub steps of the technical assessment are shown below in Table 1.

◯ HARDSHIP ANALYSIS

Hardship is taken as an indicator for the situational analysis of local drinking water services. Level of service in drinking water is based on the round-trip time required to fetch water, quality, quantity, availability and regularity. Table 2 shows the criteria for defining water service levels. This service level works as an indicator for calculating the hardship score. Based on the descending order of hardship scores, wards are classified as first, second, third and so on. Accordingly, the selected wards are prioritized at local level for selection of sites for implementing projects.

TABLE 2 Criteria for assessing the service level.

Service level (SL)	Average fetching time (min)	Quantity in liters per capita per day (lpcd)	Quality	Reliability (mon)	Continuity (hr)	Hardship ranking
Good (SL 1)	≤ 15	≥ 45	No contamination	12	≥ 6	Good
Moderate (SL 2)	> 15 but < 30	< 45 but > 25	Moderate contamination	≥ 11	≥ 5	Moderate
Poor (SL 3)	> 30 but < 45	< 25 but > 15	High contamination	≥ 10	≥ 4	Poor
Very poor (SL 4)	> 45	< 15	Very high contamination	< 10	< 4	Very poor

Hardship scores were calculated from the following mentioned formula, in numbers which vary from 0 to 300. In 3R based WUMP, hardship scores were analyzed by ward and cluster. The following formula was used for determining the hardship score of the wards of the VDC using the service level as shown above.

Hardship Score = 0 × (% of HH in SL1) + 1 × (% of HH in SL2) + 2 × (% of HH in SL3) + 3 × (% of HH in SL4

(*Source:* District Water Supply and Sanitation Development Profile, DWSS/ RWSSP Lumbini, 1993)

○ GIS BASED OPPORTUNITY MAPPING FOR A 3R APPROACH

Opportunity mapping is the tool based on GIS software ArcGIS (Version 10.1) which incorporates different secondary data such as topographic, soil, hydrology as well as primary data collected through various field surveys to analyze the feasibility of 3R interventions. Opportunity mapping can broadly be generalized into two stages, the preliminary opportunity map and the final opportunity map which are integrated into the assessment phase of the WUMP. Fig. 4 below shows GIS application in the WUMP process as the creation of opportunity maps.

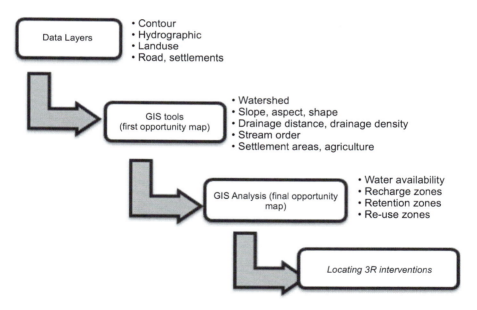

Figure 4 Flow chart resembling step wise process of GIS in 3R based WUMP.

Preliminary Opportunity map is the map composed prior to the field data collection. It consists of various secondary data of the locality, which includes watershed areas, slope, aspect, drainage density, settlement location, cultivation areas and hill-shade areas. This secondary data was used to identify potential water hardship settlement areas and locations for implementing appropriate 3R technology. Its main objective is to facilitate the field technician to identify a geographical water hardship area, available water resources in the neighborhood locality and estimate suitable 3R technologies. Using the preliminary opportunity map, the individual settlement locations were visually analyzed with reference to the hydrological network and slope of the terrain. This analysis identified the critical settlements with water hardship potential (based on visual analysis such as proximity to drainage networks, ridge based settlements, settlements located in areas having small or no hydrological basin, steep slopes and aspect where there is no or little available water sources to be used by the settlement. Furthermore, the hydrological sub basin was determined using Digital Elevation Model (DEM). These sub basins for individual hydrological network were analyzed with respect to the general terrain slope, aspect, land cover (land-use), as well as existing physical infrastructure such as roads to conclude on the suitable 3R measures for the locality. This map was aimed to help in reducing field time for the field technician as most of the information is readily available and for verification of secondary field survey data analysis using GIS.

The final opportunity map is the verified and validated preliminary opportunity map incorporating field findings and suitability analysis of 3R technologies identified by the community members. In the analysis phase of the WUMP process, GPS data of all water sources, collected during the technical assessment, were plotted with satellite imagery. The location of sources was then spatially analyzed for its topographic properties: slope, aspect and land cover data of the VDC. Based on these topographic features of the plotted water sources, different water sources conservation and improvement technology integrating 3R technologies were proposed for implementation. The final opportunity map contains the visual information of 3R interventions required in the locality. It aims to help experts to visualize feasibility of individual technologies and prioritize. This also helps the local community members to visualize the water resources in their catchment area and better understand the interventions to assist them in their sustainable management of those resources (Simonovic 2012). In addition, it provides a strong basis for the local authorities as well as central level policy makers to seek the funds for the implementation of this technology.

Identification of 3R Interventions from GIS Analysis

Different types of available 3R technologies were identified based on the information depicted by parameters of opportunity maps created using GIS. The parameters depicted on the map are chosen such that it correlates with generalized recharge and retention classifications. The parameters used to identify effective 3R interventions and locations are described below:

Slope: Slope is defined as the rate of change of elevation and it is extracted from the Digital Elevation Model (DEM). The degree of recharge of ground water table is dependent on the slope of the terrain (FAO 1968; Duchaufour 1995; Doerliger 1996; Amharref et al. 2001).

TABLE 3 Terrain slope classification (Amharref et al. 2001).

Slope (threshold)	Surface runoff	Recharge	Retention
>20%	Very high	Very weak	Very strong
6-20%	High	Weak	Strong
3-6%	Low	Strong	Weak
<3%	Very low	Very strong	Very weak

Aspect: Aspect refers to the horizontal direction to which a slope faces and it is also extracted from the DEM. In the context of Nepal – a northern hemisphere country, a south-facing slope is more open to sunlight and warm winds and is therefore generally warmer and drier due to higher levels of evapotranspiration than a north-facing slope. The water sources on northern aspect are more perennial than the southern aspect slope. Besides, the selection of 3R technologies such as gully plugging and surface ponding will be more reliable on a northern aspect slope as there will be less loss of reserved water due to evaporation than that on the southern aspect slope.

Land Cover/Land Use: Land cover/Land use has influence on the recharge of the ground water table (Musy and Soutter 1991; Dale and McLaughlin 1999). (Amharref et al. 2001) considered forest area as the high density vegetation zone, pasturage as average density, cultivated fields as weak density and bare topsoil as a very weak density vegetation zone and classified the degree of recharge accordingly. The land-use pattern can be accounted while mapping geomorphological, lithological, drainage pattern and slope maps (Singh et al. 2013b). Exploitation of natural vegetation causes problems like little scope for soil moisture storage, high rate of soil erosion (Paudel et al. 2014), declining groundwater level and shortage of drinking water (Singh et al. 2011). With this study, it can also be concluded that retention of runoff has similar behavior as recharge for the case of land use and the subsequent retention of surface runoff increases with the decrease in vegetation density. Further, other geomorphic parameters such as soil texture, land form and drainage were also preliminarily analyzed during the process as shown in Table 4.

TABLE 4 Land cover/Land use classification (Amharref et al. 2001).

Land cover	Land use	Recharge	Retention
Forest, built-up (density strong)	Coniferous, hardwood, mixed, shrub forest	Very weak	Very strong
Grass, orchard, bush (density average)	Grazing land	Weak	Strong
Cultivation (weak density)	Hill slope cultivation, valley cultivation	Strong	Weak
Barren land, sand area	sand/gravel/boulder	Very strong	Very weak

Thus, according to the location of the sources, aspect, soil type, depth, land use and slope, different appropriate 3R technologies were proposed for the improvement and conservation of the watershed area such as – small water harvesting ponds, sand dams, cutoff drains and waterways, demi lunes, recharge pond, eyebrows, small storage hill dams, small water harvesting ponds, percolation ponds, subsurface dams and contour trenches.

○ RESULT AND DISCUSSION

The WUMP process for Padukasthan VDC was successfully completed in 2013. From the information produced from GIS, the overall hydrology of the VDC was critically analyzed and appropriate solutions for preserving the catchments/watersheds were identified. The outputs of the overall analysis done in the 3R based WUMP are discussed below:

Findings on Classification of Water Sources and Decision Making for Future Projects

All of the 129 spring sources that were used as different water sources were analyzed based on their discharge and altitude. The net water budget appeared to be optimum for the present scenario with the surplus of approximately 35 m^3 per day only. However, this does not provide much flexibility in terms of water use. In the near future, it will create insufficient water supply as the gap between supply and demand is going to increase due to the positive population growth, water stress and drying up of sources (Loucks et al. 2005). Analysis reveal that the future water needs are intricate and might get daunting if no action is taken (Hanjra and Qureshi 2010). As some schemes may provide an optimal supply and others may not, insufficient water service delivery will occur with independent water schemes. Hence, the necessity of 3R based WUMP was observed for integrated management of available water resources. The sources were classified based on their discharge – litres per second (lps) and their altitude (in meters). The sources were further categorized by excluding the sources having dry season yield less than 0.1 lps as depicted by the blue shade in Fig. 5. Further, the water system analysis including the sources having a safe yield (0.9 times dry season yield) was done; as suggested by the Department of Water Supply and Sewerage's Design Guidelines for Community Based Gravity Flow Rural Water Supply Schemes (Volume 1-12). During the dry season, the discharge capacity of 75 percent of sources is less than 0.1 lps and the supply from 21 percent of sources are between 0.1-0.5 lps. Taking into account the 'potential water stress', impact of climate change and drying up of water sources in the region, considering sources with safe yield more than 0.1 lps acts as a factor of safety for incorporating climate resilient approaches in selecting sources for new projects. The reliable water sources (safe yield sources with discharge greater than 0.1 lps), also the minimum standard for supply distribution pipeline for gravity-fed system (WaterAid 2013), are generally located downstream of the settlement as shown in Fig. 7, thus causing a major number of clusters located at heights to suffer from water deficit if the lower discharge sources dry out. As the majority of the reliable water sources in wards 6 and 7 are located at an optimum height and have surplus water availability as shown in Fig. 7, they should function as major intake points for the integrated water supply system, which supply water to the entire VDC in an integrated way. Moreover, this analysis also suggests to donors or implementing agencies to focus on lift water supply schemes, rather than gravity supply schemes, by tapping the sources in valleys. The distance to source from clusters as shown in Fig. 7 and the vertical shift in height (as shown in Fig. 6, calculated from GPS data) gives the tentative figure necessary for planning the future water schemes in the zone for comparison with fetching time from a field survey. Application of low cost water lifting technology and rainwater harvesting technology are crucial to integrate livelihood into integrated water resource management (Merrey et al. 2005). Figs. 5 and 6 below highlights the classification of source based on their discharge and time to fetch water.

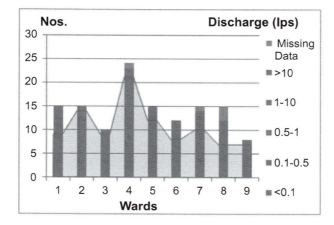

Figure 5 Summary of water sources by discharge in different wards.

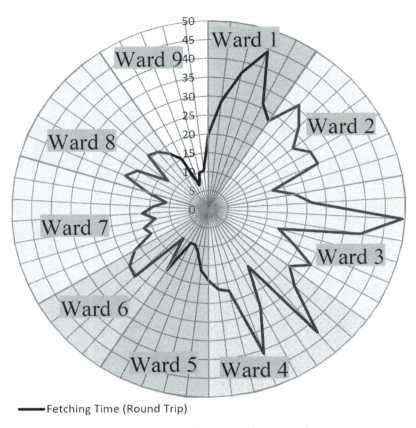

Fetching Time (Round Trip)

Figure 6 Fetching time (round trip) in different wards (in minutes).

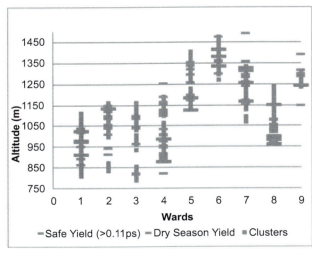

Figure 7 Altitude variation of sources with respect to clusters in different wards.

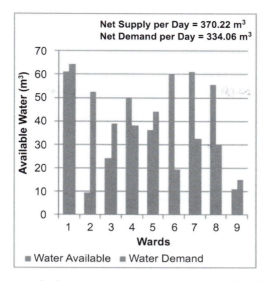

Figure 8 Net water budget in terms of supply and demand in different wards.

Findings on Hardship and Service Level

The technical analysis shows that no household in the VDC is receiving good service level (SL 1) as 46% of households have moderate service level (SL 2), 20% poor service level. (SL 3) and 34% have very poor service level (SL 4) as shown in Fig. 9. The hardship scores were calculated at the cluster level as shown in blue dots and at the ward level as shown by the red line in Fig. 10. Based on the existing drinking water service level, ward 5 has more difficulty than other wards.

Existing Service Level

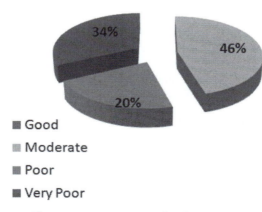

Figure 9 Existing service level in VDC.

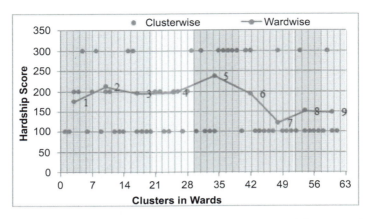

Figure 10 Hardship of existing drinking water service in different wards.

Overlapping of GIS Layers for Analysis of the Watersheds and Locating Hardship Area

Individual settlement clusters were analyzed with respect to the topographic properties, proximity of settlement to the hydrological network and its location in a hydrological sub-basin. This output of visual analysis was plotted as apotential water hardship settlement area. There were five clusters identified initially in the VDC and were marked on the Preliminary Opportunity map as shown in the GIS map (Fig. 11: Hydrological Sub-Basin with Water Hardship Area Map). Slope, Aspect and Land Cover of these individual area codes were analyzed to propose the site specific 3R Technology. GIS maps analyzing hydrological basin, DEM, Slope, aspect, land cover, hydrological sub-basin with water hardship area and spring water sources are shown in Fig. 11.

These areas were later validated in the field during the technical assessment. The GPS readings of all the existing water sources collected during this technical assessment were plotted in these maps as well as in the satellite image of the area during the report preparation phase. The topographic properties of these sources were also extracted from these preliminary opportunity maps. Based on these properties (slope, aspect, land cover), suitable 3R technology for individual sources were identified. The correlation among the findings of the preliminary opportunity map and the field data were also analyzed. It was found that the hardship area identified during the preliminary opportunity map was later confirmed by the field survey as the areas with high water scarcity. The areas experiencing water stress are those where the supply drastically falls below optimum level which cause hardship and impede development (UNDP 2007). The findings, along with the proposed 3R technology for both the water sources and the identified cluster were verified during the VDC Level planning. The outputs from the VDC level planning were incorporated to produce the final opportunity map as shown in Fig. 11. The

GIS maps of slope, hardship areas and landform for each identified hardship areas (labeled Area Code 1, 2, 3 etc.) have been analyzed in detail as shown below in Fig. 12.

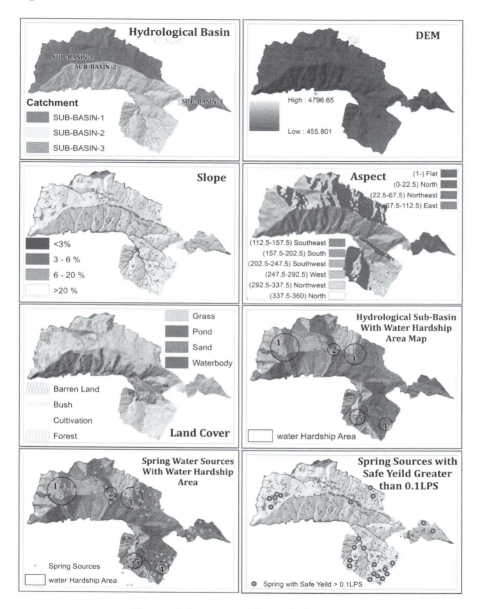

Figure 11 GIS maps of the study location.

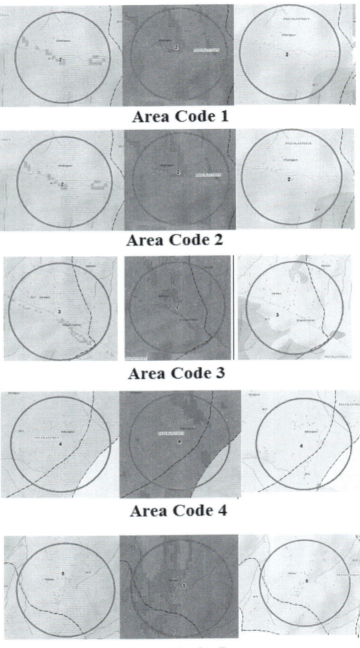

Figure 12 GIS maps of slope, hardship areas and landform for each identified hardship areas (labeled Area Code 1, 2, 3 etc.).

Area Code 1

The settlement in this area is on slope greater than 20% with some clusters on a flat slope ranging from 6-20%, north-east and west aspect with cultivation land cover. Water sources do not exist in this area. The most recommended technology would be rooftop rainwater harvesting in individual buildings and a recharge pond in the flat slope area having provision of water diversion from the road and also tied ridges in the cultivation area.

Area Code 2

The settlement of this area is on a slope greater than 20% with a limited neighborhood area on a 6-20% slope, with a north and south-west aspect and cultivation land cover. Water sources do not exist in this area. The recommended technology would be rooftop rainwater harvesting in individual buildings, retention basins and eyebrows at the area of 6-20% slope having provision of diversion of road drainage to this basin. Also a contour bund in the cultivation area is recommended.

Area Code 3

Most of the settlements in this area are along the ridge. The slope of the area is between 6-20% with some inner areas on 3-6% slope. The aspects of the area are both north and north east as it is basically a catchment divider. The land cover of the area is mostly cultivated land. The water sources do not exist in this area, as rivulets are below the settlement area. The recommended technology would be rooftop rainwater harvesting in individual buildings and a retention basin at the area of 6-20% or 3-6% slope having provision of diversion of road drainage to this basin preferably on northern aspect. Also, contour bund in the cultivation area is recommended.

Area Code 4

The settlement in this area is on a slope greater than 20% slope, with north-west aspect and cultivated land cover. The water sources do not exist in this area. The recommended technology would be rooftop rainwater harvesting in the individual buildings and a retention basin at the area of 3-6% slope having provisions of diversion of road drain to this basin. Also trapezoidal bunds in the cultivation land are recommended.

Area Code 5

Most of the settlement is on the slope greater than 20% and limited area on 6-20% slope. The aspects of the area are north and north-east. The land cover of the area is mostly cultivation land. The water sources do not exist in this area, as rivulets are below the settlement area. The recommended technology would be rooftop rainwater harvesting in individual buildings and a retention basin at the area of 6-20%. Trapezoidal bunds in the cultivation land are also recommended.

Findings on Locating 3R Infrastructures for Preserving Springshed Area

Analyzing the information from Tables 3 and 4 with the produced GIS layers, different 3R interventions (technologies) were suggested from 3R based WUMP in the final opportunity map as shown in Fig. 13. Approximately 6,040 people of 1,007 households have been indirectly benefitted by the inclusive plan. The area was found to receive a moderate amount of rainfall which increased the discharges from springs during the post monsoon period. To compare the water discharge of dry season and wet season, randomly selected sources were surveyed. The survey results revealed that the flow (discharge of water source) in rainy season is nearly 100 times greater than that in summer. This suggests there is ample of excess water in the wet season that could potentiality be retained and used in the dry spells of the year.

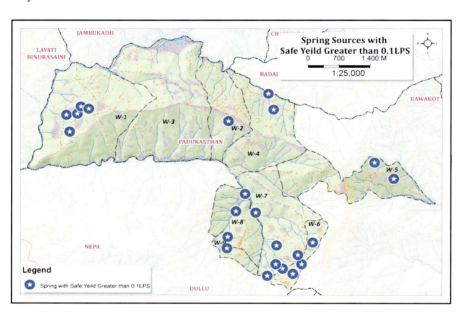

Figure 13 Proposed 3R technologies in the final opportunity map.

The identified sub projects from this WUMP have been mainstreamed into the development works of the Government of Nepal. The water supply schemes and 3R interventions of the 3R based WUMP have been implemented by the WARM-P as a project of the Department of Water Supply and Sewerage in Padukasthan. Aferro-cement lined soil cement pond (3 cum size) connected with a rain water harvesting jar is the new innovation of HELVETAS Swiss Intercooperation Nepal, Water Resources Management Programme in this VDC. Currently, 218 households have been directly benefitted through different schemes including rain water harvesting jars of 6.5 cum (90 Nos), ferro-cement lined soil cement ponds

(60 Nos), source improvement and conservation (1 Nos) and recharge ponds (3 Nos). The identified recharge interventions were plotted in the GIS layer in order to locate the infrastructures to be built in the future. One catchment restoration scheme (plantation, water retaining pits along contour line, gully control measures, eyebrows and source improvement) has been under implementation in a 5 ha area of the VDC. This will benefit an additional 110 households. The results of piloting the 3R based WUMP are promising in terms of awareness raising of the community for climate change adaptation as well as planning and implementation of schemes.

○ CONCLUSION AND RECOMMENDATION ════════════

WUMP is a participatory and transparent process which empowers marginalized groups to claim their rights to an equitable share of water within and between communities. From a gender perspective, women's main role of managing domestic water is highly considered and they get involved in all steps of the whole WUMP preparation process. Their voices and demands are incorporated in planning and implementation. GIS supported WUMP provides a clear picture of the present physical environment/scenario like catchment, slope, bifurcation ratio, land use, available water budget, settlement pattern, hardship of each settlement, etc. It also facilitates the consideration of possible hardships in the future due to the impacts of climate change on the net water budget. This helps to identify the critical watershed areas and hardship zones and prioritize accordingly the future water projects and conservation activities. This includes 3R interventions for balancing ecosystem services (water and land use) in those catchments from the early stage. Using GIS images, settlement patterns and distribution of water sources were analyzed comparatively which initially identified the water sources scarce areas which were later confirmed with the field survey. This analysis helped for prioritization of critical water supply schemes during the planning.

Different water schemes for different proposes have been identified based on demand from each cluster and resources available. Catchment areas have been assessed based on slope, aspect and land use to identify appropriate 3R interventions. Based on the location of water sources as plotted, different water sources conservation and improvement technology integrating 3R technologies have been suggested for implementation. This is crucial for the sustainability of water sources and reliability in the context of too much water and too little water due to climate change. Thus, 3R based WUMP is a pragmatic concept where technologies or interventions are assessed with local level planning which contribute a holistic hydrological basin for preserving the watersheds and sustaining the water availability. GIS mapping helped to delineate the watershed, location of water sources and their potential uses for different settlements. Moreover the mapping provided basis for the carrying out opportunity mapping. GIS was used to illustrate the spatial information in various ways to identify the different natural zones with land use and opportunities for 3R applications. The opportunity map assisted in identifying 3R application zones (ridge, hill slopes, valley foot and valley flats) encompassing the land class or existing land use.

This 3R based Water Use Master Plan of Padukasthan VDC provides the platform for annual planning and prioritization of the schemes during planning. It assists stakeholders to regularly and easily update the records of water resources and water related facilities or infrastructures. It is a strong tool for investors, donors and line agencies to cooperate and implement the water projects.

○ ACKNOWLEDGEMENT

The authors would like to acknowledge and express their appreciation to the following organizations and individuals for their contribution in preparing this paper.

- Jane Nichols, Environment Engineer (Australia),
- HELVETAS Swiss Intercooperation Nepal, Water Resources Management Programme,
- Department of Water Supply and Sewerage (DWSS) and Government of Nepal,
- Rain Foundation, Netherlands,
- Padukasthan Village Development Committee, Dailekh.

○ REFERENCES

Amharref, M., Mania, J. and Haddouchi, B. 2001. Adaptation of an Evaluation Vulnerability Method to Groundwater Pollution. First International Conference on Saltwater Intrusion and Coastal Aquifers-Monitoring, Modeling and Management. Essaouira, Morocco.

Avtar, R., Thakur, J.K., Mishra, A.K. and Kumar, P. 2012. Geospatial technique to study forest cover using ALOS/PALSAR data. pp. 139-151. *In*: Thakur, J.K., Singh, S.K., Ramanathan, A., Prasad, M.B.K. and Gossel, W. (eds.). Geospatial Techniques for Managing Environmental Resources. Springer and Capital Publication, Heidelberg, Germany.

Dale, P.F. and McLaughlin, J.D. 1999. Land Administration. Clarendon Press, Oxford.

Dent, D., Dubois, O. and Dalal-Clayton, B. 2013. Rural Planning in Developing Countries: Supporting Natural Resource Management and Sustainable Livelihoods. Routledge.

Diwakar, J. and Thakur, J. 2012. Environmental system analysis for river pollution control. Water, Air, & Soil Pollution 223(6): 3207-3218.

Doerliger, N. 1996. Advances in karst groundwater protection strategy using artificial tracer test analysis and multi-attribute vulnerability mapping (EPIK method), Ph.D. Thesis. University of Neuchâtel, Switzerland, 225 p.

Duchaufour, P. 1995. Pédologie: Sol, végétation, environnement. Masson, Paris.

Durham, B., Rinck-Pfeiffer, S. and Guendert, D. 2002. Integrated water resource management through reuse and aquifer recharge. Desalination 152: 333-338.

FAO 1968. Guidelines for Soil Description. Roma, Food and Agricultural Organisation of the United Nations.

Gupta, M. and Srivastava, P.K. 2009. Water Resource Management in Basaltic Hilly Terrain in Part of Panchmahal, Gujarat, India. International Conference on Water, Environment, Energy and Society (WEES), January 12-16, 2009, New Delhi, India.

Hanjra, M.A. and Qureshi, M.E. 2010. Global water crisis and future food security in an era of climate change. Food Policy 35(5): 365-377.

Helvetas. 2009. "Water Use Master Plan (WUMP)." Retrieved 01.08.2015, 2015, from http://assets.helvetas.org/downloads/wump.pdf.

Helvetas. 2011. Experiences of Water Use Master Plan in Nepal, Rural Water Supply in the 21st Century: Myths of the Past, Visions for the Future. 6th Rural Water Supply Network Forum. Uganda. 2011. http://www.solutionsforwater.org/wp-content/uploads/2012/01/WUMP_HELVETAS_NEPAL_RWSN-FORUM_2011.pdf.

ICIMOD, HELVETAS, et al. 2013. Natural Resource Management Approaches and Technologies in Nepal: Approach – Water use master plan. M.R. Bhatta. Kathmandu, Nepal, HELVETAS Swiss Intercooperation.

Kaburi, A.N. and Odero, P.A. 2014. Mapping and Analysis of Landcover Changes in the Upper Gucha Catchment using GIS and Remote Sensing. Proceedings of 2014 International Conference on Sustainable Research and Innovation, Volume 5, 7th-9th May 2014.

Loucks, D.P., Van Beek, E., Stedinger, J.R., Dijkman, J.P. and Villars, M.T. 2005. Water Resources Systems Planning and Management: An Introduction to Methods, Models and Applications. UNESCO, Paris.

Merrey, D.J., Drechsel, P., de Vries, F.P. and Sally, H. 2005. Integrating livelihoods into integrated water resources management: taking the integration paradigm to its logical next step for developing countries. Regional Environmental Change 5(4): 197-204.

Mukherjee, S., Sashtri, S., Gupta, M., Pant, M.K., Singh, C., Singh, S.K. et al. 2007. Integrated water resource management using remote sensing and geophysical techniques: Aravali Quartzite, Delhi, India. Journal of Environmental Hydrology Volume 15, Paper 10.

Musy, A. and Soutter, M. 1991. Physique du sol. Presses Polytechniques et Universitaires Romandes, Lausanne.

Nathanail, C.P. 2013. Decision Support Systems, Environmental. Wiley StatsRef: Statistics Reference Online.

Neupane, M., Opoku, R., Sharma, A., Adhikari, R., Thakur, J.K. and Kafle, M. 2013. Rural cold storage as a post-harvest technology system for marginalized agro-based communities in developing countries. pp. 99-112. In: Bolay, Jean-Claude, Hostettler, S. and Hazboun, E. (eds.). Technologies for Sustainable Development. Springer, Heidelberg.

Neupane, M., Thakur, J.K., Gautam, A., Dhakal, A. and Pahari, M. 2014. Arsenic aquifer sealing technology in wells: a sustainable mitigation option. Water, Air & Soil Pollution 225(11): 1-15.

Nizami, A., Hemani, M., Ara, R. and Bukhari, N. 2013. Integrated Development Planning at Village and Union Level Embarking on practical methodology. Islamabad, Pakistan, PPF, KFW.

NMIP and DWSS 2014. Nationwide Coverage and Functionality Status of Water Supply and Sanitation in Nepal. Panipokhari, Kathmandu, National Management Information Project, Department of Water Supply and Sewerage.

Padukasthan, V.D.C. 2013. Water Use Master Plan with 3R (WUMP + 3R) Padukasthan VDC., Padukasthan Village Development Committee – Dailekh.

Patel, D.P., Dholakia, M.B., Naresh, N. and Srivastava, P.K. 2012. Water harvesting structure positioning by using geo-visualization concept and prioritization of mini-watersheds through morphometric analysis in the Lower Tapi Basin. Journal of the indian society of remote sensing 40(2): 299-312.

Paudel, D., Thakur, J.K., Singh, S.K. and Srivastava, P.K. 2014. Soil characterization based on land cover heterogeneity over a tropical landscape: an integrated approach using Earth Observation datasets. Geocarto International 30(2): 1-55.

Rautanen, S. and White, P. 2013. Using every drop – experiences of good local water governance and multiple-use water services for food security in Far-western Nepal. Aquatic Procedia 1: 120-129.

Rautanen, S.-L., van Koppen, B. and Wagle, N. 2014. Community-driven multiple use water services: lessons learned by the rural village water resources management project in Nepal. Water Alternatives 7(1): 160-177.

Simonovic, S.P. 2012. Managing Water Resources: Methods and Tools for a Systems Approach. Routledge, N.Y.

Singh, P., Thakur, J.K., Kumar, S. and Singh, U.C. 2011. Assessment of land use/land cover using Geospatial Techniques in a semi-arid region of Madhya Pradesh, India. pp. 152-163. *In*: Thakur, J.K., Singh, S.K., Ramanathan, A., Prasad, M.B.K. and Gossel, W. (eds.). Geospatial Techniques for Managing Environmental Resources. Springer and Capital Publication, Heidelberg, Germany.

Singh, P., Thakur, J.K. and Kumar, S. 2013a. Delineating groundwater potential zones in a hard-rock terrain using geospatial tool. Hydrological Sciences Journal 58(1): 213-223.

Singh, P., Thakur, J.K. and Singh, U. 2013b. Morphometric analysis of Morar River Basin, Madhya Pradesh, India, using remote sensing and GIS techniques. Environmental Earth Sciences 68(7): 1967-1977.

Snellen, W. and Schrevel, A. 2004. IWRM: for sustainable use of water—50 years of experience with the concept of integrated water management: Background document to FAO/Netherlands Conference on Water for Food and Ecosystems. N. a. F. Q. Ministry of Agriculture, The Netherlands. Wageningen.

Srivastava, P.K., Han, D., Rico-Ramirez, M.A., Bray, M. and Islam, T. 2012a. Selection of classification techniques for land use/land cover change investigation. Advances in Space Research 50(9): 1250-1265.

Srivastava, P. K., Singh, S., Gupta, M., Thakur, J.K. and Mukherjee, S. 2012b. Modeling impact of land use change trajectories on groundwater quality using remote sensing and GIS. Environmental Engineering and Management Journal (in press).

Thakur, J.K., Srivastava, P.K. et al. 2012. Ecological monitoring of wetlands in semi-arid region of Konya closed Basin, Turkey. Regional Environmental Change 12(1): 133-144.

Thakur, J.K., Thakur, R.K., Ramanathan, A., Kumar, M. and Singh, S.K. 2013 Arsenic contamination of groundwater in Nepal—an overview. Water 3: 1-20.

UNDP 2007. Human Development Report 2006: Coping with water scarcity Challenge of the twenty-first century, UN-Water, FAO.

Van Steenbergen, F. and Tuinhof, A. 2010. Managing the water buffer for development and climate change adaptation. Groundwater Recharge, Retention, Reuse and Rainwater storage. Wageningen: 3R Water Secretariat.

Vyas, R. and Pandya, T.K. 2013. Extraction of hydro-geomorphologic features using satellite data for Mandsaur District, Madhya Pradesh. International Journal of Remote Sensing & Geoscience 2(3): 65-69.

WaterAid 2013. Gravity-fed schemes, Available online at www.wateraid.org/technologies.

3

CHAPTER

◇◇

Spatial Integration of Rice-based Cropping Systems for Soil and Water Quality Assessment Using Geospatial Tools and Techniques

Prem Chandra Pandey,[1] *Akhilesh Singh Raghubanshi,*[2]
Vinay Mandal,[3] *Vandana Tomar,*[4] *Swati Katiyar,*[5]
N. Ravishankar,[6] *Pavan Kumar*[5,] * and *M.S. Nathawat*[7]

ABSTRACT

Agriculture, being a backbone to of India, provides livelihood to about seventy percent of the population and from an economic point of view, from an economic perspective, contributes forty percent towards Gross National Product. High crop yield is termed as "The Green Revolution" now days. It depends on the various factors like climatic conditions, soil type, management practices and other inputs like fertilizers. Fertilizers play the role of the main contributing factor in obtaining an enviable crop yield, but the usage of high quality and quantity of fertilizers to gain more crop yield causing considerable environmental complications. Digital processing of remotely sensed sensing data opened up new opportunities for understanding synoptic and substantial changes

[1] Centre for Landscape and Climate Research, Department of Geography, University of Leicester, Leicester, UK.
[2] Institute of Environment and Sustainable Development, Banaras Hindu University, Varanasi, India.
[3] Research Associate, PDFSR Modipuram, Meerut, UP, India.
[4] Research Officer, Haryana Institute of Public Administration (HIPA), Gurgaon, India.
[5] Department of Remote Sensing, Banasthali University, Newai, Tonk, Rajasthan, India.
[6] Sr. Scientist, PDFSR, Modipuram, Meerut, UP, India.
[7] School of Life Sciences, Indira Gandhi National Open University, New Delhi, India.
* Corresponding author: pawan2607@gmail.com

in the cropping pattern. The present paper is the study of Rice Equivalent Yield (REY) and Sodium Absorption Ratio (SAR) with the effect of different fertilizer amounts like Nitrogen (N), Phosphorous (P) and Potassium (K) in the rice-based cropping pattern, i.e., Rice-Wheat, Rice-Maize, Rice-Mustard, Rice-Lytherus in Katihar and Bhagalpur districts of Bihar, India. The spatial maps of rice-based cropping with the availability of N, P and K are illustrated to determine the REY in the study area. The different soil micronutrients, i.e., Organic Carbon (OC), Electrical Conductivity (EC) and pH are also used to examine the spatial discrepancies of rice-based cropping systems productivity. Ultimately the relational pattern has been drawn an equivalent rice field and nitrate between rice equivalent yield and nitrate.

KEYWORDS: Cropping system, Fertilization Pattern, REY.

○ INTRODUCTION

Agricultural management practices in today's world are becoming technical based on the use of crop on specific soil type, amount of fertilizers and the other indicators that maintain or enhance productivity and uphold soil and water quality and catchment health. The soil testing indicators provide research results to increase the crop yield and maintain the sustainability of the soil. The indicators of sustainability have been defined as relative crop transpiration, the degree of moisture saturation, the annual change in soil water storage, soil moisture, salt concentration and its fluctuations followed by acceptable crop productivity (Agarwal and Roest 1996). The agricultural performance is based on the indicators, these are cropping intensity, cropping pattern and crop productivity. The changes in soil quality are monitored using a range of physical (soil erosion, depth, aggregation and aggregate stability, bulk density, infiltration, total and air-filled porosity, compaction, hydraulic conductivity), chemical (pH, organic C, total N, electrical conductivity and acidity, available macro- and micronutrients) and biological (microorganisms, microbial biomass and activity, respiration, mineralizable N, microbial biomass C and N, earthworm and termite biomass) characteristics (Pathak et al. 2005). An integrated approach is also needed to plan and monitor the land resources management taking a step forward to maintain sustainability and environment quality. The unmanaged practices effects soil and water management at watershed or catchment level (Waniet et al. 2003; Twomlowet et al. 2008a). These practices have negative effects on soil quality that leads to soil degradation. The negative effects can be broadly classified in such a way that the first effect is caused by soil loss by water and wind erosion (Biggelaar et al. 2004) and the second takes place due to deterioration in physical, chemical and biological properties of the soil (Poch and Martinez-Casanovas 2006).

The main problems like loss of organic matter, water logging, salinization and alkalization of the soil and the contamination of water resources are the results of physical, chemical and biological deterioration. Katyal 2003; Pathak et al. 2005 observed that the intensification of production systems without adequate investment to sustain the system, results in the loss of fertility (Carpenter 2002; Lal 1997, 2004; Biggelaar et al. 2004). Soil plays a significant role in providing agro-ecosystem supporting services, i.e., nutrient cycling and primary productivity (de Groot et al. 2002; Carpenter 2002; Waniet et al. 2005). The changes in water quality are monitored and assessed using a number of visual (color, odor, floating matter), physical (turbidity, dissolved solids, sediment load, suspended organic and inorganic materials), chemical (pH, electrical conductivity, dissolved oxygen, chemical oxygen demand, nitrate, phosphate, fluoride, pesticides and other toxic compounds, heavy metals) and biological parameters (pathogens, cyanobacteria, biomass, biological oxygen demand, phytoplankton) of water (Sahrawat et al. 2005).

GIS has emerged as a powerful tool in linkage with agro-ecological zoning and automated logic integration of bio-climate, terrain and soil resource information. With the advantage of assessing environmental changeover in a large area, remotely sensed imageries have been extensively used to acquire a wide variety of information of the Earth's surface, ranging from military applications to environmental changes detection in vegetation cover and water pollution (Boyd et al. 2002; Chang et al. 2010; Chen and Pei. 2007; Liou et al. 2001; Wang et al. 2010).

○ MATERIAL AND METHODS ══════════════════════

Study Area

Katihar and Bhagalpur districts are agriculturally dominant districts in the region of northeast lies between 25.53° – 25.24° N and 87.58° – 86.97° E (Fig. 1) and occupies an area of 3,057 and 2570 sq. km respectively. The Rivers named "Ganga" and "Mahananda" are its lifelines. Katihar and Bhagalpur are situated in the plains of the north-eastern part of Bihar and bounded by Purnea in the east, Maldah (W.B.) in the west, Kishanganj in the north and Sahebgunj (Jharkhand) in the south. The annual rainfall remains above 1148 mm under the influence of the southwest monsoon along with the mean annual temperature between 10°C and 44°C. The study area forms a part of the mid-Gangetic alluvium plain. The soils in the study area are mainly derived from the older and newer alluvium. These plain alluvial soils are light gray to dark gray in color, rather heavy and texturally fine in nature. The pH values range from neutral to acidic and the acidity of the soil gradually increases from north to south. The hilly soils are acidic with low nitrogen, medium to high potash.

Field-based Investigations

In the year 2005-06 ground truth data was collected using the handheld GPS along with district wise parts of Katihar and Bhagalpur districts. Information from agricultural fields like soil and ground water sample were collected from the district. Several ancillary dataset like administrative Boundary, other basic or statistical dataset were collected for reference purposes.

The stratified random sampling has been performed for thirty-two villages in the district. Majority (approx. 70%) of the area was composed of Rice–maize systems. The soil specimen has been sampled more comprehensively for getting more accurate fertilizer and soil amendment information. Field cultivation of rice and wheat rotation in these villages for farmer's selection was a criterion for more than ten years successively. The farmers of each experimental plot were interviewed for recording their current crop management practices, nutrient use, rice and wheat production levels during May-June 2005-06. All the surveyed farmers applied N, P and K fertilizer to both rice and wheat with considerable limits of pH, EC, OC, SAR and RSC residuals.

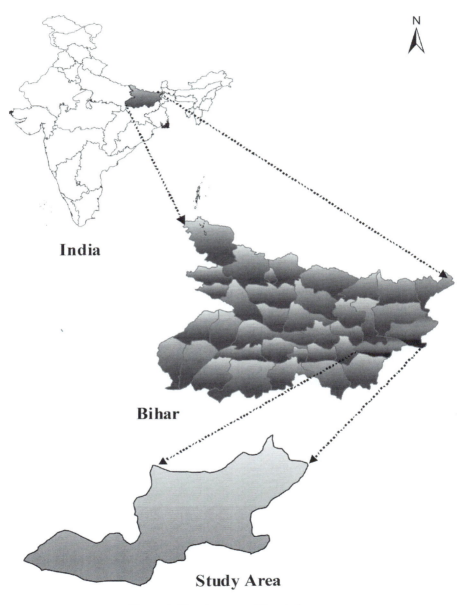

Figure 1 The site map of the study area.

Chemical Perusal of Soil Specimen

Soil samples were collected at 0 to 15 cm depth from five places in each of the 86 farmers' fields with the consideration of post wheat harvesting and before the commencement of the experiment in the year 2012-13. Soil samples collected from each field were composited and mixed; a subsample was pulverized using a wooden pestle and mortar and passed through a 100 mm sieve. Soils were analyzed for extractable N by the alkaline $KMNO_4$ method (Subbiah and Asija 1956), extractable P (0.5 M $NaHCO_3$, pH 8.5 extraction) (Olsen et al. 1954), exchangeable K (1 M NH_4OAc, pH 7.0 extraction) (Helmkeand Sparks 1996). Particle size analysis was conducted by the pipette method on initial soil samples. The N content was determined by the K *jeldahl* method using an auto analyzer. The vanadomolybdate yellow color method (Piper 1966), has been used to determine the P and total K content was determined by flame photometry.

Determination of Availability of Nitrogen (N)

Alkaline $KMNO_4$ method determines the available nitrogen in the soil with the help of alkaline potassium permanganate solution and liberated ammonia. Therefore the soil test was conducted for N (Kg ha^{-1}) (Subbiah and Asija 1956) and Nitrogen status of soil can be determined using the equation 1.

$$N_{soil} = \frac{(S - B) \times 0.00028 \times 10^6}{\text{Wt of soil sample}} \tag{1}$$

where;   S = Sample reading; B = Blank reading

Determination of Availability of OC (%) by Wet Digestion

The Organic matter in the soil oxidized with $K_2 Cr_2 O_7$ and concentrated $H_2 SO_4$ utilizing the heat of dilution of H_2SO_4. Unused $K2Cr_2O_7$ is back-titrated with $(FeSO_4.7H_2O)$ or $[FeSO_4.(NH_4)_2 6H_2O]$.

$$OC_{soil\%} = \frac{10(B - S) \times 0.003 \times 100}{\text{Wt of soil sample}} \tag{2}$$

whereas;   S = sample reading; B = Blank reading

Determination of Availability of Phosphorus (P)

The soil test consisted of Ca, Al and Fe-P quantification in the soil which indicates that Ca-P is the most dominant fraction in the soil. Organic-P fraction, was also available in a considerable amount, but not included in the determination of available P (Kgha^{-1}).

$$P_{available} = \frac{Q \times V \times 2.24 \times 10^6}{A \times S \times 10^6} \tag{3}$$

whereas; Q = quantity of P in (μ_g) on X-axis against a sample reading,
         V = volume of extracting reagent used (mL);
         A = volume of aliquot used for color development (mL);
         S = weight of soil sample taken (g).

Determination of Availability of Potassium (K)

Adjusting the instrument to zero as blank, a standard curve has been plotted against K concentrations in ($Kgha^{-1}$) by recording the flame photometer reading for each of the working standards of K.

$$K_{available} = \frac{C \times 25 \times 10^6 \times 2.24}{5 \times 10^6} \tag{4}$$

whereas; C = Concentration of K in the sample obtained on X-axis, against the reading.

Determination of Sodium Absorption Ratio (SAR)

The SAR is calculated using the amount of sodium, calcium and magnesium in the soil sample, formulated as:

$$SAR = \frac{Na^+}{\sqrt{\frac{1}{2}Ca^{2+} + Mg^{2+}}} \tag{5}$$

Determination of Residual Sodium Carbon (RSC)

By subtracting the quantity of Ca^{2+} and Mg^{2+} from the total of carbonates and bicarbonates, the residual sodium carbonate can be calculated separately in a given sample using Equation 6:

$$RSC = (CO_2 + HCO_3) - (Ca^{2+} + Mg^{2+})$$

Attainment to Rice Equivalent Yield (REY)

To calculate the equivalent rice yield, the ground truth data was compared among different sequences for easy comparison. The product is taken as individual crop was converted to REY considering the local market prices in calculation by the following formulation:

$$REY = \frac{Crop\ yield \times price\ of\ crop}{mraket\ price\ of\ rice} \tag{7}$$

○ RESULT AND DISCUSSION

The completion of the research divulges to field survey data and evaluation of rice-based cropping system in the study area. The objectives of this study entail evaluating and characterizing current cropping systems through various parameters/ indicators and suggest required diversification/intensification and analyze long-term changes in the cropping system using historical agricultural information. During the year of reporting 2014-15, we synthesized the surveyed information and analyzed the spatial integration of rice-based cropping systems productivity over Katihar and Bhagalpur districts in relation to spatial variability of fertilizer use, fertilizer availability pattern, pH, electrical conductivity (EC), Sodium absorption ratio (SAR), Residual sodium carbon and organic carbon (OC) of soil.

Rice-based cropping system is the predominant cropping system, which occupies around 70.0% (Katihar) and (54.65% approx.) under the Bhagalpur districts of the total agricultural area followed by 22.7% under Maize-based cropping system. Rice-wheat rotation is the dominant rice-based cropping system, which occupies 96% of the area. The rice-lentil and rice-lather are the minor rice-based cropping system followed in the study area. As the study area has a moderately developed irrigation facility due to the highly flooded region, it has the massive potential of fertilizer use in the region. The most commonly used fertilizers were nitrogenous, phosphatic and potassic and some micro-nutrients like Zn. The total sample survey of about 40 farmers and fertilizer use (NPK) per cropped area showed that there exists a large spatial variation of N @ 144 kgha^{-1} to 50 kgha^{-1} during Kharif season in Katihar in rice-wheat cropping system while, during rabi season, it varies from 144 kgha^{-1} into 58 kgha^{-1}. During rabi season, consumption of N, P and K is dispersal in Bhagalpur district varied from N @ 300 kgha^{-1} to 51 kgha^{-1} in rice-wheat and rice-mustard cropping system. During rabi season, it varied from 300 kgha-1 into 58 kgha-1 under the rice-based cropping system.

Impact of Micronutrient and their Application in Cropping Systems

The total fertilizer amount used also showed that per cropped area use of N was maximum @ 300 kgha^{-1} in a kharif season over Bhagalpur while available N was maximum @ 200 kgha^{-1}. Similarly, during rabi season in Bhagalpur district, N was maximum @ 300 kgha^{-1} in rice-wheat cropping systems, in Katihar use of N was maximum @ 144 kgha^{-1} in the kharif season rather than the maximum N @ 155 kgha^{-1} in rabi season. The preceding results of the multi-location research underlined the significance of improved nutrient management over farmers' fertilizer practice in enhancing crop yield under the rice-wheat cropping system. The findings are of particular significance for the intensively-cropped study area, wherein prevailing fertilization practices are skewed towards P, varied from 0 to 150 kgha^{-1} in Bhagalpur in Kharif season, while in rabi season it varied from 35 to 150 kgha^{-1}. On the other hand, Katihar district under use of P, varied from 0 to 58 kgha^{-1} in Kharif season, in comparison rabi season varied from 0 to 115 kgha^{-1}. The highest use of K was of 50 kgha^{-1} is in rice-potato over Bhagalpur districts. The use and availability of micronutrients in study areas are shown in Table 1, 2 and 3. The spatial integration of N, P and K is shown in Fig. 2.

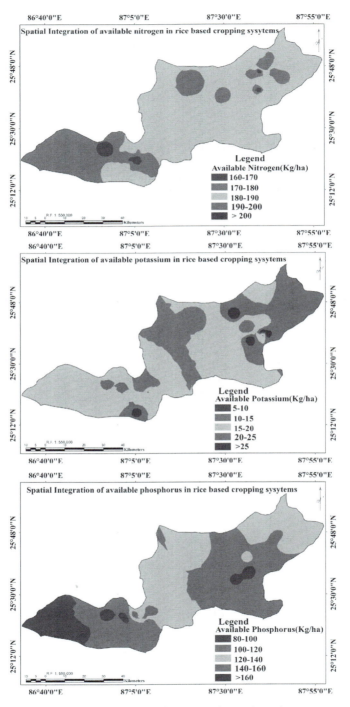

Figure 2 Spatial integration of available NPK with rice-based cropping systems.

TABLE 1 Distribution of use and available micronutrients of study area.

Dist	Crops	pH	RSC	SAR	NO_3	NK	NR	TN	PK	PR	TP	KK	KR	TK	EC	OC	AVN	AVP	AVK	REY
Bhagalpur	R-W	8.2	1.6	4.3	2.8	300	300	600	150	150	300	25	25	50	0.3	0.5	188	8	113	6.3
Bhagalpur	R-P	8.1	2.4	5.4	3.4	300	300	600	150	150	300	50	50	100	0.3	0.5	176	14	106	7.1
Bhagalpur	R-Mz	8.3	4.2	6.5	3.4	138	138	276	69	69	138	15	15	30	0.4	0.5	176	17	110	6.8
Bhagalpur	R-L	7.6	0.8	2.7	2.8	138	184	322	0	46	46	0	0	0	0.3	0.5	176	17	136	4.2
Bhagalpur	R-Mz	7.8	1.2	5.6	3.9	92	58	150	35	35	69	0	0	0	0.6	0.4	176	10	113	4.5
Bhagalpur	R-P	7.6	1.8	5.3	3.9	81	58	138	35	35	69	15	15	30	0.8	0.4	176	9	86	4.5
Bhagalpur	R-P	7.2	0.6	6.1	3.9	81	81	161	35	58	92	0	30	30	0.4	0.4	163	12	99	4.5
Bhagalpur	R-Mz	7.7	2.2	8.0	3.4	58	58	115	35	35	69	15	15	30	0.6	0.5	188	12	96	6.1
Bhagalpur	R-L	8.1	1.2	3.8	3.9	81	81	161	35	46	81	15	15	30	0.3	0.5	188	12	104	6.1
Bhagalpur	R-M	8.1	2.8	6.1	2.8	69	81	150	46	46	92	15	15	30	0.6	0.5	163	14	115	5.4
Bhagalpur	R-W	7.3	3.6	6.6	4.5	173	138	311	0	69	69	0	36	36	0.3	0.5	200	16	192	4.0
Bhagalpur	R-Ly	8.2	1.4	3.7	3.9	173	152	324	0	28	28	0	36	36	0.3	0.4	176	19	179	3.6
Bhagalpur	R-Mz	7.4	2.4	7.5	2.8	173	97	269	0	55	55	0	36	36	0.4	0.5	188	29	142	4.2
Bhagalpur	R-W	7.2	4.6	5.3	2.8	173	173	345	0	69	69	0	36	36	0.5	0.5	188	18	183	4.5
Katihar	R-Mz	7.4	0.6	7.5	2.8	58	58	115	46	115	161	30	10	40	0.2	0.5	188	13	143	7.1

Table 1 Contd.

Dist	Crops	pH	RSC	SAR	NO$_3$	NK	NR	TN	PK	PR	TP	KK	KR	TK	EC	OC	AVN	AVP	AVK	REY
Katihar	R-M	8.1	0.6	6.1	2.8	58	92	150	23	46	69	30	0	30	0.3	0.4	188	18	132	7.1
Katihar	R-Ly	8.2	2.4	5.2	3.9	92	92	184	46	46	92	30	10	40	0.4	0.5	201	18	137	7.0
Katihar	R-M	8.4	1.4	7.0	3.9	115	115	230	58	58	115	38	38	75	0.2	0.5	201	7	204	5.5
Katihar	R-W	7.3	1.8	6.6	4.5	115	115	230	58	58	115	38	38	75	0.2	0.6	213	7	183	6.3
Katihar	R-L	8.1	1.4	3.8	3.9	115	0	115	58	0	58	0	0	0	0.3	0.5	188	7	200	6.3
Katihar	R-W	8.1	0.2	4.4	3.4	115	115	230	58	20	78	38	38	75	0.2	0.6	188	14	187	7.2
Katihar	R-Ly	8.3	0.2	5.1	3.4	115	86	201	58	58	115	38	38	75	0.3	0.4	163	16	175	6.4
Katihar	R-Mz	8.5	0.2	4.7	3.9	115	115	230	58	58	115	38	38	75	0.3	0.6	188	25	139	6.9
Katihar	R-W	8.5	0.2	5.5	4.5	144	115	259	86	58	144	38	38	75	0.3	0.6	188	15	190	7.4
Katihar	R-Mz	8.4	0.6	5.1	3.9	50	40	90	20	20	40	10	10	20	0.4	0.5	176	16	179	7.9
Katihar	R-W	8.1	0.4	5.1	3.4	115	144	259	58	58	115	38	38	75	0.3	0.5	188	7	200	5.4
Katihar	R-M	8.1	0.8	6.1	2.8	86	86	173	0	72	72	0	38	38	0.2	0.5	176	10	147	5.9
Katihar	R-Ly	8.2	0.8	5.2	3.9	86	86	173	0	58	58	0	38	38	0.2	0.5	176	10	157	6.2
Katihar	R-M	8.4	0.4	3.7	3.9	86	86	173	0	72	72	0	38	38	0.2	0.4	163	9	156	6.5
Katihar	R-Ly	8.2	0.2	7.0	3.9	58	58	115	58	58	115	38	38	75	0.3	0.7	251	15	132	6.5
Katihar	R-Mz	7.4	0.2	7.5	2.8	58	58	115	58	58	115	38	38	75	0.4	0.5	188	14	145	5.9
Katihar	R-W	7.2	0.6	5.3	2.8	86	86	173	58	58	115.00	18.75	18.75	37.50	0.2	0.6	200	12	124	5.9

TABLE 2 Distribution of soil micronutrients in Katihar.

Crops	pH	RSC	SAR	NO₃	TN	TP	TK	EC	OC	AVN	AVP	AVK	REY
Rice-Wheat	7.7	0.6	5.6	3.8	103	52	28	0.24	0.54	192	11	173	6.4
Rice-Mustard	8.3	0.8	4.9	3.4	86	20	17	0.22	0.47	182	11	160	6.3
Rice-Maize	7.9	0.4	6.2	3.4	70	45	29	0.30	0.52	185	17	151	7.0
Rice-Lythrus	8.2	0.9	5.6	3.8	88	40	26	0.28	0.52	198	15	150	6.5
Rice-Lentil	8.1	1.4	3.8	3.9	115	58	0	0.34	0.45	188	7	200	6.3

TABLE 3 Distribution of soil micronutrients in Bhagalpur.

Crops	pH	RSC	SAR	NO₃	TN	TP	TK	EC	OC	AVN	AVP	AVK	REY
Rice-Wheat	7.70	2.66	5.30	3.44	294.03	115.47	35.57	0.39	0.50	187.94	13.83	145.94	5.77
Rice-Maize	7.57	3.21	5.62	3.60	235.45	81.05	34.71	0.42	0.50	187.93	15.76	164.01	5.01
Rice-Potato	7.52	3.17	5.51	3.44	249.67	87.08	35.57	0.44	0.52	191.04	16.25	171.58	5.44
Rice-Mustard	7.54	3.19	5.57	3.52	242.56	84.07	35.14	0.43	0.51	189.49	16.01	167.80	5.22
Rice-Lentil	7.53	3.18	5.54	3.48	246.11	85.57	35.36	0.43	0.52	190.26	16.13	169.69	5.33
Rice-Lythrus	8.20	1.40	6.99	3.92	324.30	27.60	36.00	0.35	0.39	175.60	18.77	179.12	3.56

Impact of pH in the Rice-based Cropping System on Soil Health and Groundwater Quality

The drainage pattern of the area indicates a dendritic pattern. Most of the agricultural farms in the study area practice the rice-wheat and rice-maize cropping systems; when pH 7.5 and 7.6 accumulated in the soil profiles the REY recorded 9.4 and 7.1 t ha^{-1} in the study area under the rice-wheat cropping system. The pH value was recorded highest 8.2 in Bhagalpur and Katihar (Table: I and III). The spatial integration of pH and REY are shown in Fig. 3.

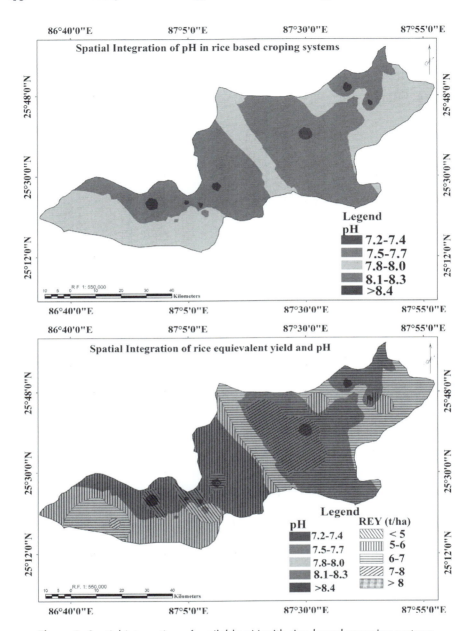

Figure 3 Spatial integration of available pH with rice-based cropping systems.

Spatial Integration EC and OC Distribution with Rice-based Cropping Systems

The average EC and OC showed that there exists a wide spatial variability under rice-based cropping system of the study area (Fig. 4). EC showed a higher value of

0.80 dSm⁻¹ in Bhagalpur and 0.15 dSm⁻¹ in Katihar rice-potato and rice-mustard cropping systems respectively. The organic carbon showed a higher value of 0.66% at Koshi Mahananda Doab region to 0.38% at Bhagalpur under rice-lytherus and rice-wheat cropping systems (Tables 1 and 4).

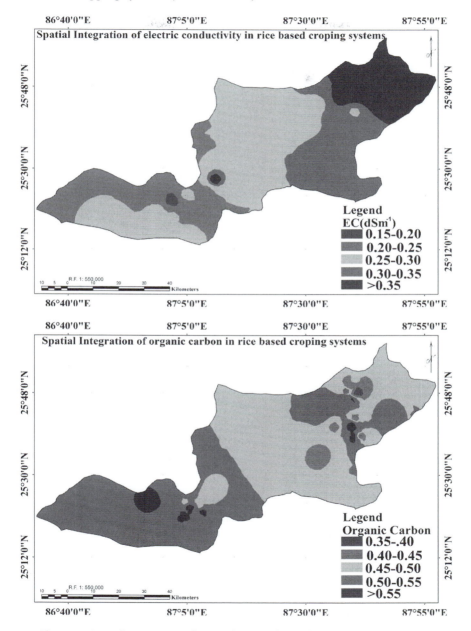

Figure 4 Spatial Integration of EC and OC with rice-based cropping systems.

Spatial Integration of Sodium Absorption Ratio Distribution with Rice-based Cropping Systems

The mean increase in rice-based cropping systems yield with SAR measurement was 7 t ha^{-1} at 6.2. SAR, 5.0 tha^{-1} at 5.6 SAR and 6.5 t ha^{-1} at 5.6 SAR (Table 2). The highest SAR response in rice at all the REY rates was recorded at 7.5 (6.3 to 7.1 tha^{-1}) at Katihar, whereas the same was lowest at 3.8 to 8.0 SAR (3.6 to 9.4 tha^{-1}) at Bhagalpur (Table 1 and Fig. 5).

Grain yields of wheat were increased with increasing rates of SAR and the yield was maximum when recorded at 9.4 kg ha-1 SAR in Bhagalpur (Fig. 1). At this SAR, the average highest response was recorded at 6.2 (7.0 t ha^{-1}) in Katihar followed by 3.8 SAR (6.3 t ha^{-1}) and 5.6 SAR (6.4 t ha^{-1}). The lowest SAR response to 3.7 was noticed at Bhagalpur (3.6 t ha^{-1}). The application of SAR at 5.5 and 6.2 produced grain yield response of 5.5 and 6.5 t ha^{-1} respectively (Tables 2 and 3).

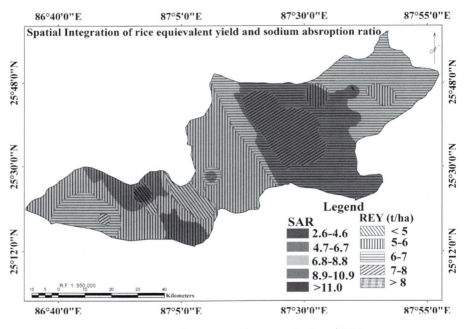

Figure 5 Spatial integration between SAR and REY.

Spatial Integration Residual Sodium Carbon Distribution with Rice-based Cropping Systems

The preceding results of the mulch-location study underlined the significance of improved nutrient management over farmers' fertilizer practice and local state recommendations, in enhancing crop yield under the rice-based cropping system.

The findings are of particular significance for the intensively-cropped study area, wherein RSC are prevailing (Fig. 6).

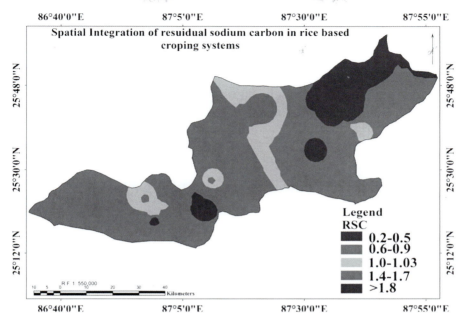

Figure 6 Spatial integration of residual sodium carbon distribution with rice-based cropping systems.

Spatial Integration Nitrate Distribution with Rice-based Cropping

To know the empirical relationship of yield variability and Nitrate (NO_3), the yields were grouped into three categories based on the standard deviation (Group I (high) – < 10.0 t/ha, Group II (normal) – 5 – 10.0 t/ha and Group III (low) – > 5 t/ha). It is found that the higher rice equivalent yield of the rice-based cropping system may be due to the higher use of N and with the available Nitrate (NO_3). (Table 4 and Figs. 7 and 8).

TABLE 4 Distribution of REY Groups and soil micronutrients of the study area.

REY Group	pH	RSC	SAR	NO_3	TN	TP	TK	EC	OC	AVN	AVP	AVK	REY
> 5	7.7	2.4	6.5	3.6	223.3	63.5	26.4	0.4	0.5	178.0	15.4	133.7	4.2
5-7	8.0	1.4	5.5	3.5	208.6	105.1	46.6	0.3	0.5	189.8	12.1	146.8	6.2
< 7	8.0	0.8	5.7	3.5	223.5	137.7	50.6	0.3	0.5	183.4	14.8	151.7	7.5

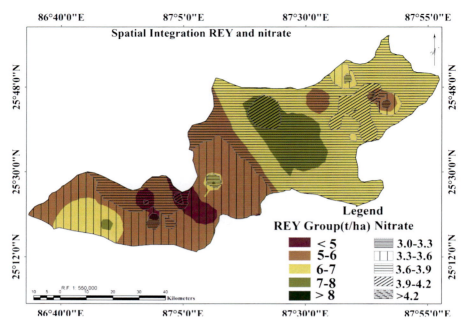

Figure 7 Spatial distribution of Nitrate and REY.

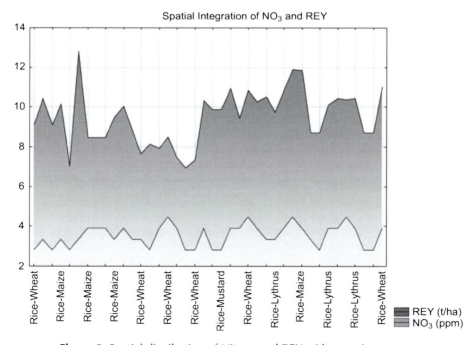

Figure 8 Spatial distribution of Nitrate and REY with cropping system.

◯ CONCLUSION

The study of the multi-site crop field and micronutrients give an impression of improved nutrient management over farmer's fertilizer practice and local state recommendations that can enhance crop yield and net returns estimates of the rice-wheat cropping system. The outcomes indicate that the intensively-cropped IGP prevalent fertilization practices are skewed towards N that indicates the high use of N fertilizer. Further, optimization of fertilizer P and K input build upon crop demands and native supplies from soil nutrient pools at dislocations efficiently increases P and K, increase agronomic efficiency and recovery efficiency. An increase in P use efficiency consequential to the inclusion of K in fertilizer planning and vice-versa implies the necessity of both P and K application in adequate amounts to achieve volumetric efficiencies.

◯ ACKNOWLEDGMENT

We are highly thankful to the Space Application Center (SAC), Ahmedabad, India for providing the ancillary dataset. Authors would like to thanks also the Image Processing Lab, Department of Remote Sensing, Banasthali Vidyapith, Rajasthan, India for providing all the necessary support.

◯ REFERENCES

Agarwal, M.C. and Roest, C.J.W. 1996. Towards improved water management in Haryana state, Final report of the Indo-Dutch operational research project on hydrological studies, CCS Haryana Agricultural University, India.

Biggelaar, C.d., Lal, R., Wiebe, K. and Breneman, V. 2004. The global impact of soil erosion on productivity. I: Absolute and relative erosion-induced yield losses. Advances in Agronomy 81: 1-48.

Boyd, D.S., Foody, G.M. and Ripple, W.J. 2002. Evaluation of approaches for forest cover estimation in the Pacific Northwest, USA, using remote sensing. Applied Geography 22: 375-392.

Carpenter, S.R. 2002. Ecological futures: building an ecology of the long now. Ecology 83: 2069-2083.

Chang, T.-Y., Liou, Y.-A., Lin, C.-Y., Liu, C.-S. and Wang, Y.-C. 2010. Evaluation of surface heat fluxes in Chiayi plain of Taiwan by remotely sensed data. International Journal Remote Sensing 31(14): 3885-3898, DOI: 10.1080/01431161.2010.483481.

Chen, J., Lin, H. and Pei, Z. 2007. Application of ENVISAT ASAR data in mapping rice crop growth in Southern China. IEEE Geoscience Remote Sensing Letters 4: 431-435.

de Groot, R.S., Wilson, M. and Boumans, R. 2002. A typology for the description, classification and valuation of ecosystem functions, goods and services. Ecological Economics 41: 393-408.

Helmke, P.A. and Sparks, D.L. 1996. Lithium, sodium, potassium, rubidium and cesium. pp. 551-574. *In*: Sparks, D.L., Page, A.L., Helmke, P.A. and Loeppert, R.H. (eds.). Methods of Soil Analysis. Part 3. Chemical Methods. SSSA Book Series no. 5. SSSA, Madison, WI.

Katyal, J.C. 2003. Soil fertility management—a key to prevent desertification. Journal of the Indian Society of Soil Science 51: 378-387.

Lal, R. 1997. Degradation and resilience of soils. Philosophical Transactions of Royal Society, London B 352: 997-1010.

Lal, R. 2004. Soil quality indicators in industrialized and developing countries—similarities and differences. pp. 297-313. *In*: Schjonning, P., Elmholt, S. and Christensen, B.T. (eds.). Managing Soil Quality: Challenges in Modern Agriculture. CAB (Commonwealth Agricultural Bureau) International, Wallingford, UK.

Liou, Y.-A., Liu, S.F. and Wang, W.J. 2001. Retrieving soil moisture from simulated brightness temperatures by a neural network. IEEE Geoscience Remote Sensing Letters 39: 1662-1673.

Olsen, S.R., Cole, C.V., Watanabe, F.S. and Dean, L.A. 1954. Estimation of available phosphorus in soils by extraction with sodium bicarbonate. USDA Circ. 939, USDA, Washington, DC.

Pathak, P., Sahrawat, K.L., Rego, T.J. and Wani, S.P., 2005. Measurable biophysical indicators for impact assessment: changes in soil quality. pp. 53-74. *In*: Shiferaw, B., Freeman, H.A. and Swinton, S.M. (eds.). Natural Resource Management in Agriculture: Methods for Assessing Economic and Environmental Impacts. CAB International, Wallingford, UK.

Piper, C.S. 1966. Soil and Plant Analysis. University of Adelaide Press, Adelaide, Australia.

Poch, R.M. and Martinez-Casanovas, J.A. 2006. Degradation. pp. 375-378. *In*: Lal, R. (ed.). Encyclopedia of Soil Science, 2nd edition. Taylor and Francis, Philadelphia, PA, USA.

Sahrawat, K.L., Padmaja, K.V., Pathak, P. and Wani, S.P. 2005. Measurable biophysical indicators for impact assessment: changes in water availability and quality. pp. 75-96. *In*: Shiferaw, B., Freeman, H.A. and Swinton, S.M. (eds.). Natural Resource Management in Agriculture: Methods for Assessing Economic and Environmental Impacts. CAB International, Wallingford, UK.

Subbiah, B.V. and Asija, G.L. 1956. A rapid procedure for the determination of available nitrogen in soils. Current Science 25: 259-260.

Twomlow, S., Love, D. and Walker, S. 2008a. The nexus between integrated natural resources management and integrated water resources management in southern Africa. Physics and Chemistry of the Earth 33: 889-898.

Wang, Y.-C., Chang, T.-Y., Liou, Y.-A. and Ziegler, A. 2010. Terrain correction for increased estimation accuracy of evapotranspiration in a mountainous watershed. IEEE Geoscience Remote Sensing Letters 7(2): 352-356, DOI: 10.1109/LGRS.2009.2035138.

Wani, S.P., Pathak, P., Jangawad, L.S., Eswaran, H. and Singh, P. 2003. Improved management of Vertisols in the semi-arid tropics for increased productivity and soil carbon sequestration. Soil Use and Management 19: 217-222.

Wani, S.P., Singh, P., Dwivedi, R.S., Navalgund, R.R. and Ramakrishna, A. 2005. Biophysical indicators of agro-ecosystem services and methods for monitoring the impacts of NRM technologies at different scales. pp. 97-123. *In*: Shiferaw, B., Freeman, H.A. and Swinton, S.M. (eds.). Natural Resource Management in Agriculture: Methods for Assessing Economic and Environmental Impacts. CAB International, Wallingford, UK.

4
CHAPTER

◇◇

A Geographic Information System (GIS) Based Assessment of Hydropower Potential within the Upper Indus Basin Pakistan

Abdul Moiz,[1] *Muhammad Saadoon Basraa,*[1] *Muhammad Ali,*[1]
Muhammad Ali Shamim[2,*] and *Usman Ali Naeem*[1]

ABSTRACT

The increasing energy demand, due to an increasing population and the resulting strain on our traditional energy sources, has proved the importance of renewable energy sources as a prerequisite for sustainable economic growth. Hydropower is one such source that is both renewable and sustainable in addition to being environment-friendly. Over the last few years, power shortages have been experienced in almost all parts of Pakistan due to the relatively large difference between demand and supply. The aim of this study is to assess the hydropower potential of the Upper Indus Basin Pakistan using Shuttle Radar Topography Mission (SRTM) based Digital Elevation Model (DEM) along with observed stream flow records in ArcGIS environment. Using the Spatial Analyst tool, a number of locations were identified on the main Indus River for development of run-of-river hydropower plants. Results from this study could prove to be quite helpful in making precise decisions in favor of a sustainable hydropower development and finally alleviate the energy crisis being faced today by Pakistan. The study also indicates the usefulness of using GIS in carrying out spatial interpolation and consequently, determination of hydropower potential.

[1] Department of Civil Engineering, University of Engineering and Technology, Taxila, Pakistan.
[2] Department of Civil Engineering, Bursa Orhangazi University, Bursa, Turkey.
* Corresponding author: muhammad.shamim@bou.edu.tr

KEYWORDS: Geographic Information system, Hydropower potential, Inverse Distance Weighting, Digital Elevation Model.

○ INTRODUCTION

Power derived from the potential and kinetic energy of falling and running water i.e. hydropower, has long been used as a renewable source of energy. Hydropower is one of the oldest methods of producing power. The energy from moving water is used to run the turbines connected to a generator in turn generating electricity, more commonly known as hydroelectricity. Today, hydropower is one of the mostly widely-used renewable resources as 15% of the total power being produced comes from hydropower. A survey of the world energy resources shows that hydropower production is on the rise as 2,286 TWH of hydropower was generated in 1993 which increased to 3229 by the year 2011, which is expected to increase to 3,826 by the year 2020. Nonetheless, a large demand for energy is still met by burning fossil fuels such as coal, oil, natural gas etc. But due to the serious environmental hazards associated with these sources and the depleting fossil fuel reserves, these sources cannot be relied upon as a sustainable source of energy. According to the U.S. Energy Information Administration (2012), electricity generation factors contributed 32% in total greenhouse gas (GHG) emissions in the year 2012, over 70% of which comes from the burning of fossil fuels, mostly coal and natural gas. Due to these adverse effects, the world today is moving towards the development of more sustainable and eco-friendly energy sources, of which hydroelectric power is one.

Pakistan relies largely upon hydroelectric power for meeting its energy requirements. In 2013, the country generated 36% of its electricity from oil, 29% from natural gas, 29% from hydropower and 5% from nuclear (U.S. Energy Information Administration 2014). The country has a total power generation capacity of about 13,800 MW against an average demand of 17,000 MW. In other words, the country suffers from a shortfall of around 3,200 MW (NTDC, 2014). This shortfall has been the cause of severe power outages throughout the country, hampering the country's economic growth. This calls for the identification of potential hydropower development sites in order to meet the energy deficit, being a sustainable and economical solution to the country's energy crisis.

Over the year, Geographic Information Systems (GIS) have proven to be an efficient tool in the identification of potential sites for hydropower development (Ballance et al. 2000). Such tools can supplement or in some cases completely replace the "on ground" surveys, necessary for identifying suitable sites for hydropower development, due to the topographic inaccessibility. Different studies have been conducted in various parts of the globe on the application of GIS for identifying potential sites for hydropower development. Ballance et al. (2000) performed a GIS based analysis for identifying micro and macro scale hydropower potential in South Africa. Babu et al. (2010) suggested a customization in ArcGIS Explorer 900 framework to create an intuitive interface for 3D visualization of identified potential hydropower sites in Alaknanda and Bhagirathi river valleys of Uttrakhand, India. Larentis et al. (2010) investigated the use of a GIS-based computational program

'*Hydrospot*' to identify hydropower sites, based on remote sensing and regional streamflow data, in an automated environment. Buehler (2011) discussed the design of an ArcGIS toolset to estimate the flow duration curves (FDCs) at locations where data does not exist, in order to identify the most suitable site for small hydroelectric power development in the Dominican Republic. Jha (2011) worked out the run-of-river type hydropower potential of Nepal by incorporating GIS and a hydropower model. Several researchers reviewed various tools that have been used over the years for a preliminary or prefeasibility study of the site for the development of small hydropower plants (Punys et al. 2011). Wakeyama and Ehara (2011) investigated the renewable energy potential in Hokkaido, Northern-Tohoku Area and Tokyo Metropolitan areas of Japan by incorporating both, hydropower and solar energy resources. Feizizadeh and Haslauer (2012) made use of local topographic, monthly evaporation and precipitation data in a GIS-based environment for evaluating the hydropower potential of Tabriz Basin, Iran. A number of other similar works also emphasize the importance of the GIS in evaluating hydropower potential and identifying suitable sites for their development (Hall et al. 2012; Gergeľová et al. 2013; Wali 2013; Abebe 2014).

Pakistan has vast reserves of untapped hydropower resources in the Northern Region that includes the Upper Indus Basin (UIB); however, owing to the difficult terrain of the area, ground surveys to identify potential hydropower sites would be relatively expensive and time consuming too. Therefore, this study would be helpful in the sense that it will provide a basis for evaluating the hydropower potential associated with a particular site located in the Upper Indus Basin, Northern Areas of Pakistan. All this will be performed in a GIS environment using a SRTM 90 m Digital Elevation Model (DEM) and observed streamflow datasets. The study primarily focuses on hydropower potential assessment along the main Indus River, within Pakistani boundaries i.e. starting from Kharmong upto Tarbela (Fig. 1).

○ STUDY AREA AND DATASETS USED

The study area is located between the longitudes of 72°15′00″E to 77°50′00″E and latitudes of 33°50′00″N to 37°05′00″N which is a portion of the Upper Indus Basin (UIB), falling within the boundaries of Pakistan (Fig. 1) and having an area of 128,040 km². Hindukush-Karakorum-Himalaya (HKH) mountain ranges constitute a large part of the basin. This study mainly focusses on the Upper part of Indus basin which lies upstream of Tarbela reservoir, a multipurpose reservoir built on the main Indus River. Along the way, the Indus River is joined by many small and large tributaries, most prominent of which are Shyok River, Shigar River, Hunza River, Gilgit River and Astore River. Streamflow is subject to extreme seasonal variability with snowmelt and glacier runoff being the primary contributors in the higher parts of the basin (Ali and Boer 2007). Bookhagen and Burbank (2010) estimated that more than 60% of the annual runoff in the basin is attributable to snowmelt in HKH region. The river is also fed by runoff generated by rainfall, mostly in the lower parts of the basin. The region has several mountain peaks exceeding 7000 m above mean sea level (AMSL) and possesses the greatest perennial glacial ice area outside the Polar Regions of over 22,000 km² (Sharif et al. 2013). The climate of the basin is very sensitive to altitude. The mean elevation of the basin is 4750 m AMSL of which 60% area is above an elevation of 4500 m AMSL and 15% above 4750 m AMSL (Tahir 2011). Annual precipitation typically varies from 100 to 200 mm at low altitudes and sometimes exceeds up to 600 mm at 4400 m AMSL (Cramer 1993). The basin is marked with low temperatures with mean monthly temperature going below freezing from October to March, normally in areas with an elevation of above 3000 m AMSL (Archer and Fowler 2004). The extreme weather patterns and difficult terrain are the primary reasons for the sparse population in the region.

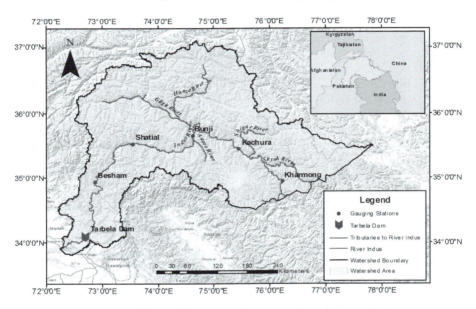

Figure 1 Location of the study area in Pakistan, its boundaries and river system.

This study makes use of the observed streamflow and topographic data for evaluating the hydropower potential in the Upper Indus Basin. The observed streamflow data, at daily intervals, for a period of 16 years i.e. 1991-2006, was acquired from Water and Power Development Authority (WAPDA) for the six stream gauging stations (Fig. 1) namely, Kharmong, Kachura, Bunji, Shatial, Besham and Tarbela located within the study area. Average annual flows were computed using this data at each of the gauging stations using Inverse Distance Weighting (IDW), incorporated within ArcGIS environment. The topographic details were obtained in the form of a Digital Elevation Model (DEM) from NASA's Shuttle Radar Topography Mission (SRTM) 90 m digital elevation data from which the study area was delineated using ArcGIS (Fig. 2).

Methodology

The adopted methodology consisted of generation of stream network and catchment delineation using ArcGIS, a geographic information system, in conjunction with

ArcHydro Tools extension. ArcHydro Tools was also consequently used for spatial interpolation of streamflows along the main Indus River using Inverse Distance Weighting, at intervals of 5 km, starting from Kharmong and moving towards Tarbela. Spatial interpolation of the stream flows was carried out using ArcGIS, that seamlessly integrates various different types of spatial information and makes possible their efficient manipulation to generate useful spatially distributed results. After the spatial interpolation of streamflows, hydropower potential of each of the interpolated points was computed. Details of the methodology are elaborated below.

Stream Network Generation

As a first step, the SRTM 90 m digital elevation data obtained from National Aeronautics and Space Administration's (NASA) Shuttle Radar Topography Mission (SRTM) in the form of four tiles, covering the whole of Upper Indus basin, located in Pakistani territory. All tiles were merged together using the "Mosaic To New Raster" function incorporated within ArcGIS. Sinks were then filled in the grid to eliminate the possibility of the water getting trapped in a cell by using the "Fill Sinks" function. The direction of flow was then computed using the "Flow Direction" function. "Flow Accumulation" was used to compute the accumulated number of cells upstream of a cell, for each cell in the input grid. The flow accumulation grid generated was then used as an input to the "Stream Definition" function, which generates a stream gird by assigning a value of "1" to all the cells in the input layer that was a value greater than a given threshold. For the purpose of this study the threshold was set at 1% of the maximum flow accumulation. The previous two layers generated, were then input to the "Stream Segmentation" function of the ArcHydro Tools to create a grid of stream segments, each with its unique identification, which may be classified as a head segment or a segment between two segment junctions. Finally, the stream network was converted to a vector using the "Drainage Line Processing" function with stream segmentation and flow direction grids as inputs.

Catchment Delineation

The next step was to delineate the catchment (study) area and make a separate Digital Elevation Model (DEM) of it using SRTM tiles within ArcGIS. Catchment delineation makes use of some of the layers generated as a part of stream network generation, performed previously. The first step towards Catchment Delineation was the "Catchment Grid Delineation" which required flow direction grid and stream segmentation grid as its inputs. The Catchment Grid Delineation function classifies the entire area under consideration into a number of catchments based on the stream segments which drain into that area. The grid was then converted to vector map by using the "Catchment Polygon Processing" function. Using the vector map generated in the previous step, the aggregated upstream catchments were created by running the "Adjoint Catchment Processing" function. The layer thus generated basically represents the whole upstream area draining to the inlet point of each catchment. Finally, "Batch Point Generation" and "Batch Watershed Delineation" functions of ArcGIS were used to delineate the study area from amongst the merged tile i.e. Upper Indus Basin was delineated upto Tarbela. The final delineated DEM of the Upper Indus Basin using the above mentioned procedure is shown in Fig. 2.

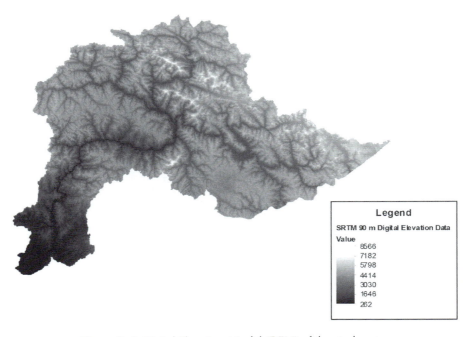

Figure 2 A Digital Elevation Model (DEM) of the study area.

Streamflow Interpolation using Inverse Distance Weighting (IDW)

The selected six gauging stations were then located within the study area i.e. on the main Indus River using their co-ordinates (Latitude; Longitude) and were assigned an average annual runoff (m³/sec or cumecs) as the attribute value. Runoff was interpolated over the entire catchment area under consideration by using Inverse Distance Weighting (IDW) technique in ArcGIS. IDW is a multivariate interpolation technique that uses weighted average of attribute values at given points to assign values to points with unknown attributes, expressed mathematically in equation (1) (Chen and Liu 2012).

$$P_i = \frac{\sum_{j=1}^{G} \frac{P_j}{D_{ij}^n}}{\sum_{j=1}^{G} \frac{1}{D_{ij}^n}} \tag{1}$$

where,

P_i = is the value of interpolated point at location i

P_j = the value measured at sampled location j

D_{ij} = the distance from i to j

G = the number of sampled locations

n = the inverse-distance weighting power

The value of *n*, in effect, controls the region of influence of each of the sampled locations. As *n* increases, the region of influence decreases until, in the limit, it becomes the area which is closer to point *i* than to any other. When *n* is set equal to zero, the method is identical to simply averaging the sampled values. In this case however, the inverse-distance weighting power was set to 2, as normally used (Zhu and Jia 2004).

Evaluation of Hydropower Potential

Points were then constructed along the entire length of 780.4 km along the River Indus, starting from Kharmong, where River Indus enters the international boundaries of Pakistan, to just downstream of Tarbela dam, at an interval of 5 km, these being the points at which the hydropower potential was to be evaluated. The runoff and elevation at these points were extracted from the DEM and interpolated runoff layer, using the "Extract Multi Values to Points" (Fig. 3). Head at each point was then evaluated by subtracting the elevation at each point from that of the immediately previous point (Fig. 4) as explainable by the following equation,

$$H_N = h_{N-1} - h_N \tag{2}$$

where,

h_N = elevation of the point under consideration

h_{N-1} = elevation of the immediate upstream point

H_N = Head in terms of height of water column at the point under consideration

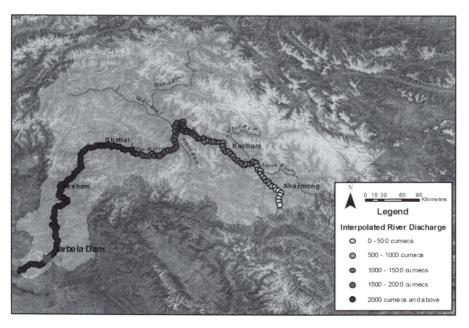

Figure 3 Discharge Interpolate by Inverse Distance Weighting (IDW) technique for Indus River.

The calculation of the theoretical hydropower potential was based on the following basic equation (Lehner et al. 2012), a schematic of which is shown in Fig. 4.

$$\text{Power output} = P = \rho Q g H \tag{3}$$

where, P = Power output in watts

ρ = Density of water = 1000 kg/m³

Q = River discharge (m³/s)

g = Acceleration due to gravity = 9.81 m³/s

H = Available head (m)

The density of water has been assumed to be constant and the acceleration due to gravity is a constant. The only two variables in equation (3) are river discharge and available head, which vary along the length of the river and have already been evaluated using ArcGIS.

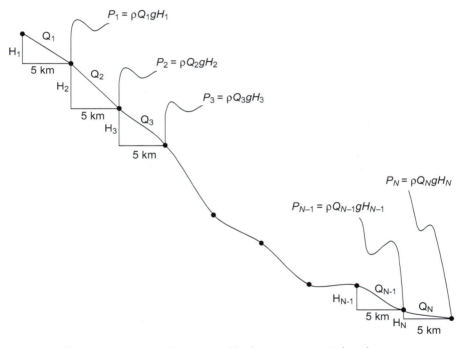

Figure 4 Schematic illustration of hydropower potential evaluation.

○ RESULTS AND DISCUSSION

Fig. 5 shows the main results of the study by providing a map of hydropower potential along the main Indus River. Additionally, locations with the highest hydropower potential are tabulated in Table 1.

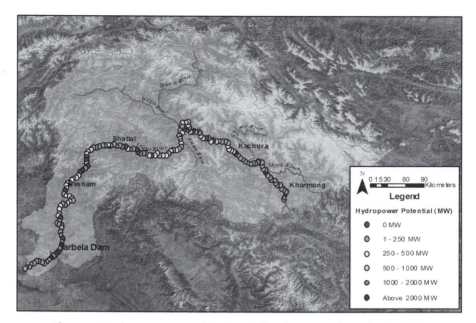

Figure 5 Hydropower potential along Indus river as evaluated using GIS.

At certain locations, the available head '*H*' was coming out to be negative, thereby indicating that the location has no hydropower potential. These locations have been highlighted in 'red' in Fig. 5. The values highlighted in green correspond to a hydropower potential greater than 1000 MW and these points are more likely to be suitable for the development of a run-of-river type hydropower scheme.

From Fig. 5, one can also identify that the hydropower potential is low towards the upstream of the river where it enters Pakistan, followed by a subsequent increase as it is joined by several tributaries towards downstream. Fig. 5 also shows a large potential near the junction of the Gilgit, Hunza and Astore Rivers with Indus in the vicinity of Bunji with values as high as 1687.9 and 1930.7 MW and an average potential of 479 MW in the Kachura-Bunji reach, much of which is concentrated near the downstream end of the reach. A significant potential for hydropower generation is also observed midway in the reach from Shatial to Besham in excess of 1000 MW (Fig. 5) with an average of 307 MW. Points that fall in the Bunji – Shatial reach exhibit a moderate power potential ranging upto 398 MW (Fig. 5 and Table 1), probably because the reach has a flatter slope as compared to the rest of the reaches, resulting in a smaller available head. The hydropower potential is low near the upstream end of the river where it enters Pakistan, followed by a subsequent increase as it is joined by several tributaries downstream. A fair amount of hydropower potential is exhibited by the areas in the vicinity of Kharmong with the maximum potential as low as 221 MW due to the low average annual runoff carried by the River Indus when it enters Pakistan. A largest hydropower potential is observed near Tarbela Dam, roughly 2200 MW, which seems consistent with the fact that Tarbela Dam is the largest earthen dam in the world and is the primary source of electrical power in the country. Top 10 hydropower potential sites are shown in Table 1.

TABLE 1 Sites with highest Hydropower potential along the Indus River.

Longitude	Latitude	Elevation above mean sea level (AMSL) (m)	Elevation above mean sea level (AMSL) of upstream end (m)	Head (m)	Average annual discharge (m³/s)	Hydropower potential (MW)	Hydroelectricity (×10⁶ KWh/yr)	Reach
72°40'35" E	34°4'6" N	332	430	98	2204.70	2119.6	18567.3	Besham-Tarbela
74°44'36" E	35°44'7" N	1498	1603	105	1874.39	1930.7	16913.1	Kachura-Bunji
74°54'9" E	35°41'45" N	1652	1750	98	1755.72	1687.9	14786.2	Kachura-Bunji
73°11'10" E	35°14'34" N	706	769	63	2254.79	1393.5	12207.3	Shatial-Besham
73°24'1" E	35°30'57" N	867	919	52	2152.79	1098.2	9620.1	Shatial-Besham
74°39'21" E	35°49'39" N	1314	1367	53	1874.95	974.8	8539.6	Kachura-Bunji
74°46'8" E	35°42'45" N	1603	1655	52	1865.78	951.8	8337.5	Kachura-Bunji
74°57'41" E	35°37'57" N	1788	1841	53	1669.11	867.8	7602.1	Kachura-Bunji
75°9'52" E	35°35'36" N	1869	1934	65	1355.52	864.4	7571.7	Kachura-Bunji
73°12'6" E	35°17'42" N	768	807	39	2225.88	851.6	7460.0	Shatial-Besham

○ CONCLUSIONS

This paper presents a very efficient approach for evaluating the hydropower potential associated with a catchment area. It also demonstrates that ArcGIS along with ArcHydro toolset is a highly effective, efficient and remarkably user-friendly software for rapid identification of potential hydropower sites. This is mainly for run-of-river development, which could have otherwise been a very cumbersome job, not to mention the susceptibility of the results to errors due to the lack of topographical data. It can also be concluded that the Upper Indus Basin possesses an abundant potential for hydropower generation, of the order of 40,711.4 MW from the point where it enters into Pakistani territory, up to Tarbela. The highest potential was found to be in the Kachura-Bunji reach, being a total of 15720.9 MW. As far as other reaches are concerned, the hydropower potential upstream of Kharmong up to the Pakistani boundary was 174.6 MW; Kharmong-Kachura 2496.7 MW; Bunji-Shatial 6146.9 MW; Shatial-Beshama 8732.5 MW and finally for Besham-Tarbla reach, the total hydropower potential was of the order of 7439.8 MW. As far as potential for a single location is concerned, the maximum power potential was found to be in the Besham-Tarbela region, of the order of 2119.6 MW. This shows that the Northern Areas of Pakistan have tremendous hydropower potential, which is still untapped. Similar studies should also be conducted for the smaller sub-catchments of Indus basin in order to develop a spatially distributed map identifying potential areas for small and large scale hydropower developments e.g. for Shatial, Gilgit, Astore, Hunza catchments. This could prove to be a significant milestone for future development of Pakistan in general and its Northern Areas in particular. It is believed that this study can definitely help the decision-making authorities like the Water and Power Development Authority (WAPDA) of Pakistan to make quick decisions in the best interest of the country.

○ ACKNOWLEDGEMENT

The authors would like to express the deepest appreciation to the Water and Power Development Authority (WAPDA) for sharing the daily runoff data with us in the best interest of this study, without which all this could not have been made possible.

○ REFERENCES

Abebe, N. 2014. Feasibility study of small hydropower schemes in Giba and Worie Subbasins of Tekeze River, Ethiopia. Journal of Energy Technologies and Policy 4(8): 8-17.

Ali, K.F. and Boer, D.H.D. 2007. Spatial patterns and variation of suspended sediment yield in the Upper Indus River Basin, northern Pakistan. Journal of Hydrology 334(3-4): 368-387.

Archer, D.R. and Fowler, H.J. 2004. Spatial and temporal variations in precipitation in the Upper Indus Basin, global teleconnections and hydrological implications. Hydrology and Earth System Sciences 8(1): 47-61.

Babu, D.G., Pandey, K., Mathew, J., Kumar, P. and Sharma, K.P. 2010. Geospatial Gateway for Potential Hydropower Sites 3D Visualization, 11th ESRI India User Conference 21-22 April, 2010, Delhi India.

Ballance, A., Stephenson, D., Chapman, R.A. and Muller, J. 2000. A geographic information systems analysis of hydropower potential in South Africa. Journal of Hydroinformatics 2(4): 247-254.

Bookhagen, B. and Burbank, D.W. 2010. Toward a complete Himalayan hydrological budget: Spatiotemporal distribution of snowmelt and rainfall and their impact on river discharge. Journal of geophysical research, 115(F3).

Buehler, B.D. 2011. Analyzing the Potential for Small Hydroelectric Power Installment in the Dominican Republic, Brigham Young University, 72 pp.

Chen, F.W. and Liu, C.W. 2012. Estimation of the spatial rainfall distribution using inverse distance weighting (IDW) in the middle of Taiwan, Paddy and Water Environment 10(3): 209-222.

Cramer, T. 1993. Climatological investigations in Bagrot Valley. *In*: Cultural Area Karakoram Newsletter. 3, 19-22, Tubingen.

Feizizadeh, B. and Haslauer, E.M. 2012. GIS-based procedures of hydropower potential for Tabriz basin, Iran, GI_Forum 2012, Salzburg, Austria.

Gergeľová, M., Kuzevičová, Ž. and Kuzevič, Š. 2013. A GIS based assessment of hydropower potential in Hornád basin. Acta Montanistica Slovaca 18(2): 91-100.

Hall, D.G., Verdin, K.L. and Lee, R.D. 2012. Assessment of Natural Stream Sites for Hydroelectric Dams in the Pacific Northwest Region. Idaho National Laboratory, Idaho Falls, Idaho 83415.

Jha, R. 2011. Total run-of-river type hydropower potential of Nepal. Journal of Water, Energy and Environment 7: 8-13.

Larentis, D.G., Collischonn, W., Olivera, F. and Tucci, C.E.M. 2010. GIS-based procedures for hydropower potential spotting. Energy 35(10): 4237-4243.

Lehner, B., Czisch, G. and Vassolo, S. 2012. Europe's hydropower potential today and in the future, [online], http://www.usf.uni-kassel.de/ftp/dokumente/kwws/5/ew_8_ hydropower_low.pdf. National Transmission and Despatch Company (NTDC). 2014. Power System Statistics. World Energy Resources. 2013 Survey: Summary. World Energy Council.

Punys, P., Dumbrauskas, A., Kvaraciejus, A. and Vyciene, G. 2011. Tools for small hydropower plant resource planning and development: areview of technology and applications. Energies 4: 1258-1277.

Sharif, M., Archer, D.R., Fowler, H.J. and Forsythe, N. 2013. Trends in timing and magnitude of flow in the Upper Indus Basin. Hydrology and Earth System Sciences 17(4):1503-1516.

Tahir, A.A. 2011. Impact of climate change on the snow covers and glaciers in the Upper Indus River Basin and its consequences on the water reservoirs (Tarbela Dam) – Pakistan, Universite Montpellier 2 Sciences et Techniques Du Languedoc.

U.S. Energy Information Administration. 2012. Electricity Explained – Basics.

U.S. Energy Information Administration. 2014. Pakistan: Country Analysis Note.

Wakeyama, T. and Ehara, S. 2011. Estimation of renewable energy potential and use: A case study of Hokkaido; Northern-Tohoku Area and Tokyo Metropolitan; Japan, World Renewable Energy Congress – Linköping, Sweden, pp. 3090-3097.

Wali, U.G. 2013. Estimating hydropower potential of an ungauged stream. International Journal of Emerging Technology and Advanced Engineering 3(11): 592-600.

Zhu, H.Y. and Jia, S.F. 2004. Uncertainty in the spatial interpolation of rainfall data. Progress in Geography 23(2): 34-42.

5

CHAPTER

◇◇◇

Flood Risk Assessment for Kota Tinggi, Johor, Malaysia

Tze-Huey Tam,[1] Ab Latif Ibrahim,[1],* Muhammad Zulkarnain Abd Rahman,[2] Ngai-Weng Chan[3] and Iain Woodhouse[4]

ABSTRACT

Flood risk maps are a vital tool to provide various valuable information for reducing flood damage and spatial planning purposes. Only the flood risk map provides information on consequences of flooding, which can be used as a primary tool to initiate a holistic flood risk management project. Geospatial data such as satellite image and LiDAR data, as well as hydrological data, are used to produce flood risk maps by integrating one and two dimensional hydraulic models. Risk is a production of hazard and vulnerability. Flood risk is expressed as economic loss and a flood risk zone map is produced in both quantitative and qualitative form. Three different depth damage functions were adopted to estimate the flood damage for three different physical elements. The average flood damage for residential areas is RM 350/m² (USD 95/m²), RM 200/m² (USD 54/m²) and RM 100/m² (USD 27/m²) using United States, The Netherlands and Malaysia damage functions. This study successfully produces flood risk but further study is needed for improve the results by taking into consideration epistemic uncertainties and develop a stage damage function tailored to studied area.

KEYWORDS: Flood risk assessment, hazard, flood damage, depth damage function, flood risk.

[1] Geoscience and Digital Earth Centre, Research Institute of Sustainability and Environment, Universiti Teknologi Malaysia, Johor Bahru, Johor, Malaysia.
[2] TropicalMap, Faculty of Geoinformation & Real Estate, Universiti Teknologi Malaysia, Johor Bahru, Johor, Malaysia.
[3] School of Humanities, Universiti Sains Malaysia, Penang, Malaysia.
[4] School of Geosciences, The University of Edinburgh, Edinburgh, UK.
* Corresponding author: ablatif@utm.my

○ INTRODUCTION

Flood is the most damaging natural disaster in Malaysia in terms of population affected, socio-economic impact and hazardous extent (Sulaiman 2007). The country had experienced many major floods since 1920. The Department of Irrigation and Drainage (DID) estimated that total flood prone areas is approximately 9 per cent of the land area of Malaysia (29,800 km²) affecting over 4.82 million people (i.e. 22% of the country's population). The average annual flood damage has been estimated at RM 0.3 billion (USD ≈ 900 Million). Kelantan is one of the 13 states of Malaysia and is greatly affected by the monsoon floods during the last decades. According to EM-DAT*, the flood event that occurred in December 2006 and January 2007 in Johor state was considered as the most costly flood event in which it is reported that USD 0.5 billion of flood damage occurred and affected 237,000 people. This event also claimed 18 lives.

The main contributing factors for the occurrence of floods in Malaysia are monsoon and convective effects which resulted in a large concentration of runoff, greatly exceeding river capacity. An increase in the concentration of runoff also attributed to the development in the floodplain which leads to an increase of flood risk in the floodplain (Caddis et al. 2012). However, the severe flood event in 2006/2007 is due to an unprecedented rainfall in which accumulated daily rainfall for 4 days continuously result in more than 250 mm of rain (Shafie 2009).

In Malaysia, the flood management system hinges on structural measures (e.g. construction of dams, dikes, levees, river deepening). Mitigating flood damage and due to such measures inevitably leads to economic benefits in terms of generating jobs, boosting the economy and producing huge returns (Chan 2005, 2012). To date, Malaysia is undergoing a programme to produce a nationwide Flood Hazard Map (FHM) (Tanaka and Kuribayashi 2010), which is one of the non-structural measures emphasized and developed by the Department of Irrigation and Drainage (DID), Malaysia (Omar 2014). The methodology and the materials used by DID for generating the flood hazard map are explained by Khalil and Lu (2014) and Omar (2014). Khalil and Lu (2014) also calculated the annual average damage for the selected land use class, viz. residential area, commercial area, industrial area, institutional, transportation, agricultural and aquaculture. Flood maps can also deliver different information that is required by the end user for their purposes and use, e.g. increasing awareness among people at risk, providing information for urban land-use planning and helping to assess the feasibility of structural and non-structural flood control measures.

To date, studies related to flood mapping have been carried out in Malaysia by using hydrologic models, hydraulic models, remote sensing-based models, or a combination of hydrological, hydraulic and remote sensing models (Pradhan 2009; Julien et al. 2010; Kia et al. 2012; Ghani et al. 2012; Mah et al. 2011). However, none of these studies focus on assessment of flood risk at a given area. This is

---

* D. Guha-Sapir, R. Below, Ph. Hoyois – EM-DAT: International Disaster Database – www.emdat.be – Université Catholique de Louvain – Brussels – Belgium. This database is maintained by Centre for Research on the Epidemiology of Disasters (CRED) with the support of the World Health Organization (WHO) and the Belgian Government.

because of the data of vulnerability is very limited and difficult to present in GIS formats. Among the available flood maps there are only flood risk maps that deliver the information on the consequence of flooding associated with a given probability of occurrence (De Moel et al. 2009).

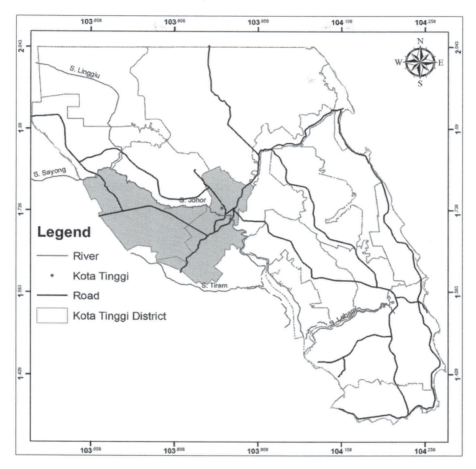

Figure 1 Location site of the study area, Kota Tinggi.

In order to overcome this gap, the objective of this study is to provide a simple but technically detailed approach of applied flood risk analysis. We apply this method in Kota Tinggi (Fig. 1) that experienced severe flood events in 2006/2007. In this study, we assess expected damage on a subset of land-use classes for different floods caused by Johor River of Kota Tinggi that is unprotected from river flooding. To assess flood risk of the study area, we do not present any new approach of risk assessment. Instead, we use existing methods of hydrodynamic modelling, geospatial tools and existing hydrological data in order to provide a complete flood risk assessment chain (i.e. assessment of hazards to development of risk maps). However, we also assess the other common parameters of flood hazards, i.e. duration and rise of the water level, in order to increase awareness of people at risk by contributing to a comprehensive understanding of their flood hazard level.

○ MATERIALS AND METHODS

Study Area

Kota Tinggi is an important town in Johor state, Malaysia. It is located approximately 40 kilometers northeast of Johor Bahru. Kota Tinggi district consists of an area of 3,500 km² (364,399 hectares) divided into sixsub-districts (Fig. 1). Urbanization in this area is growing rapidly with a focus on agricultural activities and housing development with a population of more than 200,000 people. The average elevation of Kota Tinggi is 6 meters above mean sea level.

The total length of Sungai Johor is approximately 123 km long. The Johor River inlet starts from Gemuruh Mountain and flows over the south-eastern part of Johor before reaching the Straits of Johor. The watershed has an irregular shape with an area of 2,636 km². The maximum length and breadth of the watershed are 80 km and 45 km respectively. Approximately 60% of the watershed is highland with maximum height of 366 m above mean sea level. The rest of the watershed is made up of lowlands and swampy areas.

The highland area is generally covered by tropical forest. In the south, a major portion of the forest had been cleared and planted with oil palm and rubber. The highland areas are covered in granite soil consisting of fine to coarse sand and clay. The catchment area generally receives an average annual precipitation of 2,500 mm while the mean annual discharge measured at Rantau Panjang was 37.7 m³/s. The major tributaries are Sayong River, Linggiu River, Tiram River and Lebam Rivers. Kota Tinggi is the only town located in this basin.

Johor River originates from its source of Linggiu River and Sayong River in the upstream before merging into Johor River and flows down Southeast to Johor Straits. Downstream major tributaries are Tiram River and Lebam River. Most of the downstream area is covered with wetlands and swampy area. Generally, Johor River is a meandering river with most of the areas prone to having a neck or chute cut-off with high sedimentation ranging from suspended to mixed as well as bedload sedimentation along the reaches.

Data Used

The data used for the simulating flood event included a Landsat 5 TM image, 1-m spatial resolution of the Digital Elevation Model (DEM) derived from LiDAR data, designed hydrological data at boundary (water level and discharge) and river profiles. Except for the Landsat image, which is freely available over the internet, the rest of the data are provided by DID. Each of this data will be explained in the following section in term of its sources and characteristics.

Methodology

In this study, we developed a risk-based model for Kota Tinggi, which follow the risk concept of UNISDR (2013)* in which disaster risk is a function of hazard (i.e. hazardous event associated with probability of occurrence and its characteristics, such as depth, extent, velocity, etc.), exposure (elements at risk, e.g. buildings, houses, people, agricultural) and vulnerability (what is the extent of damage or loss of the elements at risk susceptible to a flooding). Therefore, in this study, the method of assessment of flood risk consists of four steps: (i) hazard assessment; (ii) exposure assessment; (iii) vulnerability assessment; and (iv) risk assessment. Each of the stepsis explained in the following sub-section.

Hazard Assessment: Flood hazard is defined as the exceedance probability of potentially damaging flood situations in a given area and within a specified period of time (Merz et al. 2007). The calculation of probability of occurrence are carried out using historical hydrological data in order to predict the future flood event with specific return periods at a given area. Furthermore, flood hazard can also be determined based on its characteristics, such as inundation depth, inundation extent, flow velocity, rise of water level etc. However, some characteristics are difficult to quantify into measurement units, viz. contamination of flood water and debris flow (Merz et al. 2007).

Before undertaking hazard assessment, a flood simulation with different return periods should be produced. The data used for flood simulation consists of hydrological data (stream flow and water level), cross section, Manning's roughness coefficient and the Digital Elevation Model (DEM) data. Hydrological data was provided by DID with five return periods (i.e. 2, 25, 50, 100 and 200-yr return period). The severe flood event occurred in December of 2006 and January of 2007 was considered over 100-year return period (Shafie 2009). In this situation, we predicted that extreme rainfall due to climate change will be occurring in the future. Therefore, a 200-yr return period flood event was simulated to assess the expected flood risk. Cross sections were also obtained from DID. Six land use classes, i.e. built environment, grassland, oil palm, rubber, vacant area and water body were classified from the Landsat TM 5 image captured on 4 May 2005 using a supervised classification technique with the Maximum Likelihood (ML) method. The satellite image was downloaded at USGS website (http://glovis.usgs.gov/). The built environment consists of residential, commercial and industrial areas. The aimis to perform this classification technique on the satellite image in order to obtain Manning's roughness coefficient value of the land use classes. In this study, we adopted and modified the initial Manning value of Memarian et al. (2012) to apply our study area. This value is one of the main inputs to hydrodynamic modeling where it acts as a surface friction parameter to influence the movement of flood water. However, the spatial resolution of DEM was downgraded, using the cubic resampling technique, to 30 meters in order to match the resolution of the satellite

---

* ISDR (2013) Global Assessment Report on Disaster Risk Reduction, United Nations, Geneva, Switzerland. Access at: http://www.preventionweb.net/english/hyogo/gar/2013/en/home/download. html

image. Therefore, every spatial cell contains a value of elevation and a value of Manning.

The flood scenarios with different return periods at the study area was simulated by using 1D2D hydrodynamic modeling, called SOBEK. The outputs of the simulation are the inundation depth and inundation extent. On the basis of these two outputs, we derived other hazard parameters, i.e. rise of water level, flood impulse and duration. We also adopted a hazard rating (HR) for people (Defra 2006):

$$HR = h(v + 0.5) + df \qquad (1)$$

where h is flood depth (in meters), v is velocity (in meter/second) and df is the debris factor (0-2 score). The degree of flood hazard for people is low if HR < 0.75, moderate if 0.75 < HR < 1.5, significant if 1.5 < HR < 2.5 and extreme if HR > 2.5.

Exposure Assessment: This assessment focuses on identification of elements at risk in different flood scenarios. In this paper, the land use classes of Kota Tinggi district in GIS formats was provided by Kota Tinggi District Council. The provided dataset contains three crucial attributes, which are classified as Class I (Current), Class II (Activity) and Class III (Activity 2) on the basis of the land use classification code, e.g. for the case of residential, Class I is residence; Class II is planned housing; Class III is one the of non-residential strata, new village, felda village, felda housing, estate and government staff housing. The land use classes in our study area consists of water body, industry, utilities and infrastructure, institutional and community facilities, residence, beach, transportation, business and services, agriculture, vacant area and open space and recreation. However, only residence, industry and business and services land use classes were be assessed for expected damage.

The exposure of different elements to flooding is identified by overlaying the land use classes of the study area with flood extent generated from hydrodynamic modeling. However, not all the elements located on the floodplain will suffer damage and loss.

Vulnerability Assessment: Vulnerability assessment determines how likely the elements at risk are to suffer damage and loss (Schanze 2006). The vulnerability of elements to a flooding depends upon two factors: i.e. exposure and susceptibility. Susceptibility is usually determined by stage-damage functions. Such a function gives the degree of damage if an element is flooded. These functions estimate damage as a function of hazard parameters such as velocity and depth. Kreibich et al. (2009) conducted a study to investigate the impact of flow velocity in determining flood damage. They concluded that flow velocity is not recommended for use in flood damage modeling. However, the hazard rating used in this study takes velocity into consideration for assessing risk to people. Therefore, we take flow velocity into consideration for assessing flood risk. In this study, three depth-damage functions derived from the Netherlands, United States and Malaysia were used for calculating the damage factors of affected elements at risk. The value of vulnerability (damage factor) is in the range of 0 to 1.

Risk Assessment: Flood risk is a result of a consequence of a hazardous event with aprobability exceeding some threshold. In this study, flood risk is defined as the potential damage associated with a flood event, expressed as economic losses. The potential damage of different land use types is calculated as follows. Firstly, the damage of one unit property is calculated as the damage factor multiplied by property value. Secondary, the average damage of a land use-type is calculated by summing up of the damage units divided by the total affected area. Lastly, the average potential of one property can be estimated as the average damage of a land use type multiplied by the damage factor, which expressed as RM/m^2 (USD/m^2), is the standard unit for the flood risk map.

○ RESULTS AND DISCUSSION

For hazard assessment, two flood hazard maps (flood depth map and flood velocity map) were developed using hydrodynamic modeling. On the basis of these two types of flood maps, we derived another four types of flood maps, i.e. maximum rising of water level, duration, maximum flood impulse and a flood hazard (Fig. 2). The development of these types can determine the degree of flood hazard with different situations. This can help to increase the awareness among the local people regarding a flood event. The flow direction of Johor River in this figure is from left to right. These maps reveal that the extent of flooded area on the left bank is larger than the right bank. This is because the land elevation at the right bank is relatively higher. As a consequence, the area on the left bank experiences high hazard in terms of flood depth and people living in that area experience significant risk in terms of hazard risk. Based on simulation, a flood scenario with a 200 year return period is expected to last for 2-3 days or 3-4 days at Kota Tinggi, which show agreement with the true flood event that occurred on 2006/2007 (Shafie 2009). Table 1 illustrates a summary of the total flooded area in different return periods at the study area.

TABLE 1 Inundated area and non-inundated area at different flood scenarios.

Return period	Inundated area (km²)	Non-inundated area (km²)	Total of study area (km²)
2-yr	1.09	10.67	11.76
25-yr	3.78	7.98	11.76
50-yr	4.64	7.12	11.76
100-yr	5.38	6.38	11.76
200-yr	5.61	6.15	11.76

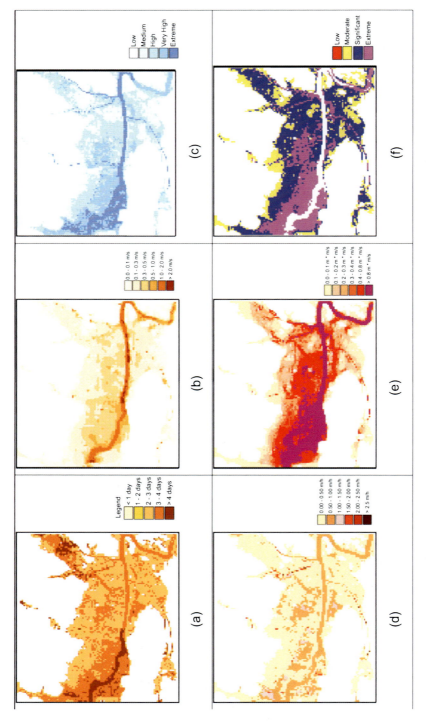

Figure 2 Flood hazard maps of 200-year return period: (a) Duration; (b) Maximum flood velocity; (c) Maximum flood depth; (d) Maximum rising of water level; (e) Maximum impulse; and (f) Hazard rating.

By overlaying the flood extent map with the structural plan, we can identify the specific elements being flooded in different return periods, shown in Table 2. Among the flooded elements, some elements, i.e. electricity substation and hospital categories are critical facilities, defined as facilities providing essential community functions, where loss of function during a flood is considered of critical importance. Furthermore, welfare homes and kindergartens are also flooded in different return periods in which these two age groups, children (age less than 7) and older citizens (age above 65), are considered highly susceptible to floods. However, local authorities can identify those community halls as safe from being flooded so that they can be used as refugee centers.

Among the three different adopted vulnerability curves (Table 3), only the vulnerability curve from the United States takes basement damage into consideration. This curve was not applied in this study for determining percent damage as local houses in our study area do not have basements. An overestimation of percent damage would therefore be expected. Fig. 3 shows the vulnerability maps for housing based on the vulnerability curve adopted from The Netherlands and from Malaysia, respectively.

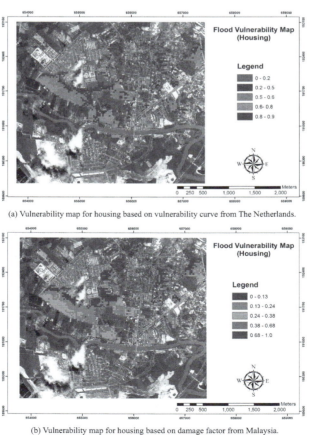

(a) Vulnerability map for housing based on vulnerability curve from The Netherlands.

(b) Vulnerability map for housing based on damage factor from Malaysia.

Figure 3 Vulnerability map at Kota Tinggi.

TABLE 2 Element being affected in different flood scenarios.

Element at risk	Return periods				
	2-yr	25-yr	50-yr	100-yr	200-yr
Electricity supply (*electrical substation & primary substation*)	–	11	12	13	15
Planned industry (*light*)	1	4	4	4	4
No planned industry (*light & medium*)	2	5	9	8	10
Religions (*temple, surau, gereja & mosque*)	–	9	9	10	11
Government use/badan berkanun (*government office/government agency*)	1	4	6	6	6
Safety (*fire station*)	–	2	6	6	6
Health (*hospital*)	–	1	1	1	1
Other community facilities (*hall & community hall*)	–	5	6	7	7
Education (*kindergarten & primary school*)	–	7	15	15	16
Planned business (*retail, services and miscellaneous*)	18	789	881	917	939
Planned housing (*housing non strata new village*)	124	2,388	2,982	3,370	3,515
No planned housing (*village traditions*)	2	12	12	12	12
Welfare homes (*elderly home, home society/association*)	–	2	2	2	2

TABLE 3 Adopted vulnerability curves applied in study area.

Land use	United States	The Netherlands	Malaysia
Housing	✓	✓	✓
Commercial area	✗	✓	✗
Industrial area	✗	✓	✓

Once inundation maps were generated and elements that are exposed to floods were identified, vulnerability assessment was then carried out as explained earlier. Fig. 3 shows the damage factors of housing by using a vulnerability curve from the Netherlands and from Malaysia in which damage factors of the Netherlands is higher than Malaysia. Note that the local data (property price, for instance) has a large impact on the development of the vulnerability curve. Some assumptions were made during the assessment of flood vulnerability. The adopted vulnerability curve were applied to all types of housing, all types of commercial shops and all types of factory.

After assessing inundation and vulnerability, flood risk maps (in term of expected damage) were developed for three land use classes on the basis of flood depth, which is considered here as the main source of damage. Expected damage values were classified in five defined classes (from 'very low' to 'very high') as shown in Fig. 4. Table 4 shows the expected damage values for land use classes. For the housing damage, the results are based on the vulnerability curve derived from United States is significantly higher than the results based on the vulnerability curve derived from the Netherlands and Malaysia. This is because the vulnerability curve from United States takes the basement of a house into consideration. As a result, this may lead to an overestimation of damage. However, overestimation may also be attributed to derived vulnerability curves from other countries (Budiyono et al. 2014). They strongly suggest that more attention needs to be paid to selection, development and testing of the vulnerability curve and it should consist of local information used to estimate the economic exposure value.

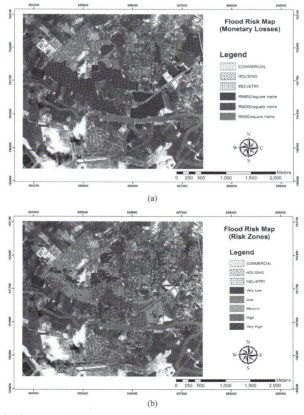

Figure 4 Flood risk map in 200-year return period: (a) Economic losses; and (b) risk zones.

TABLE 4 Expected damage.

Land use	United States	The Netherlands	Malaysia
Housing	RM 350/m²	RM 200/m²	RM 100/m²
Commercial area	✗	RM 800/m²	✗
Industrial area	✗	RM 30/m²	RM 10/m²

○ CONCLUSION

In this study, flood risk assessment of Kota Tinggi was carried out through a sequence of four steps: (1) hazard assessment, (2) exposure assessment, (3) vulnerability assessment and (4) risk assessment. Flood depth maps and flood velocity maps were developed using 1D2D hydrodynamic modeling (SOBEK) with hydrological data, cross section, Manning's value and DEM as model inputs. On the basis of these two types of flood maps, four different flood hazard maps were also developed. Identification of flooded elements in different return periods was carried out from which some critical facilities were highlighted. These facilities provide essential functions to the community. Vulnerability analysis for different land use classes was carried out using vulnerability curves from United States, the Netherlands and Malaysia.

In this study, we have developed a flood risk assessment model for Kota Tinggi providing important information and methodology development, but this work should nonetheless be considered as an interim step towards a more comprehensive flood risk assessment because of the many limitations required to be addressed in the future. The limitations included: no assessment of the social and environmental flood risk; the spatial dimension is not fully factored into the flood risk assessment model; and uncertainty from hydrological data or knowledge of flood damage estimates are neglected. All these problems should be taken into consideration for future study.

○ ACKNOWLEDGEMENTS

We would like to thanks to the following agencies for providing data to this study; Department of Irrigation and Drainage, Kota Tinggi District Council, Valuation and Property Services Department Johor Bahru, U.S Geological Survey for providing data to this study. Special thanks to Mr. Giedrius for providing HAZUS damage functions.

○ REFERENCES

Budiyono, Y., Aerts, J., Brinkman, J., Marfai, M. and Ward, P. 2014. Flood risk assessment for delta mega-cities: a case study of Jakarta. Natural Hazards 75: 389-413.

Caddis, B., Nielsen, C., Hong, W., Anun Tahir, P. and Yenn Teo, F. 2012. Guidelines for floodplain development – a Malaysian case study. International Journal of River Basin Management. 10: 161-170.

Chan, N.W. 2005. Sustainable management of rivers in Malaysia: involving all stakeholders. International Journal of River Basin Management 3: 147-162.

Chan, N.W. 2012. Managing urban rivers and water quality in Malaysia for sustainable water resources. International Journal of Water Resources Development 28: 343-354.

De Moel, H., Van Alphen, J. and Aerts, J.C.J.H. 2009. Flood maps in Europe – methods, availability and use. Natural Hazards and Earth System Sciences. 9: 289-301.

Defra 2006. Flood risks to people: Phase 2 FG2321/TR1.

Ghani, A.A., Chang, C.K., Leow, C.S. and Zakaria, N.A. 2012. Sungai Pahang digital flood mapping: 2007 flood. International Journal of River Basin Management 10: 139-148.

Julien, P., Ghani, A., Zakaria, N., Abdullah, R. and Chang, C. 2010. Case study: flood mitigation of the Muda River, Malaysia Journal of Hydraulic Engineering. 136: 251-261.

Khalil, M.K. and Lu, M. 2014. Non-structural Approach in Mitigating Flood Damage-Utilisation of Flood Hazard Map. Proc. 13th International Conference on Urban Drainage. Sarawak, Malaysia.

Kia, M., Pirasteh, S., Pradhan, B., Mahmud, A., Sulaiman, W. and Moradi, A. 2012. An artificial neural network model for flood simulation using GIS: Johor River Basin, Malaysia. Environmental Earth Sciences 67: 251-264.

Kreibich, H., Piroth, K., Seifert, I., Maiwald, H., Kunert, U., Schwarz, J., Merz, B. and Thieken, A.H. 2009. Is flow velocity a significant parameter in flood damage modelling? Natural Hazards and Earth System Sciences 9:1679-1692.

Mah, D., Putuhena, F. and Lai, S. 2011. Modelling the flood vulnerability of deltaic Kuching City, Malaysia. Natural Hazards 58: 865-875.

Memarian, H., Balasundram, S.K., Talib, J., Sung, C.T.B., Sood, A.M., Abbaspour, K.C. and Haghizadeh, A. 2012. Hydrologic analysis of a tropical watershed using KINEROS2. Environment Asia. 5: 84-93.

Merz, B., Thieken, A.H. and Gocht, M. 2007. Flood risk mapping at the local scale: concepts and challenges. pp. 231-251. *In*: Begum, S., Stive, M.F. and Hall, J. (eds.). Flood Risk Management in Europe. Springer, AA Dordrecht, The Nertherlands.

Omar, B.Z.C. 2014. Flood Hazard Maps: An Update. Proc. Malaysia Water Resources Management (MYWRM). Putrajaya, Malaysia.

Pradhan, B. 2009. Flood susceptible mapping and risk area delineation using logistic regression, GIS and remote sensing. Journal of Spatial Hydrology 9: 1-18.

Schanze, J. 2006. Flood risk management – a basic framework. pp. 1-20. *In*: Schanze, J., Zeman, E. and Marsalek, J. (eds.). Flood Risk Management: Hazards, Vulnerability and Mitigation Measures. Springer, Netherlands.

Shafie, A. 2009. Extreme Flood Event: A Case Study on Floods of 2006 and 2007 in Johor, Malaysia. Msc. Thesis, Colorado State University, Fort Collins, Colorado.

Sulaiman, A.H. 2007. Flood and Drought Management in Malaysia. Proc. National Seminar on Socio-Economic Impacts of Extreme Weather and Climate Change. Putrajaya, Malaysia.

Tanaka, S. and Kuribayashi, D. 2010. Progress Report on Flood Hazard Mapping in Asian Countries.

CHAPTER

Delineation and Zonation of Flood Prone Area Using Geo-hydrological Parameters: A Case Study of Lower Ghaghara River Valley

Rajesh Kumar,[1,]* *Sanjay Kumar*[2] and *Prem Chandra Pandey*[3]

ABSTRACT

Zonation of flood prone area on the basis of flood magnitude and geomorphic features denotes the assessment of a river valley in terms of susceptibility of floods. The Landsat and Resourcesat-1 satellite images (1977, 1990 and 2008) and Landsat MSS derived inundated area (1982) and Dartmouth flood atlas (1998, 2005 and 2008) have been used for identification of the geomorphic units and flood prone area of the lower Ghaghara river valley. The magnitude of floods in the above mentioned years has been defined on the basis of their areal extent and duration of flooding. The flood prone area along the Ghaghara River (Ayodhya-Turtipar stretch) has been divided into three flood zones using overlay of flood layers of different magnitude and geomorphic features in the GIS environment. On the basis of geomorphic features, flood magnitude and frequency, vulnerability of hamlets in each flood zone has been assessed. The channel belt and active floodplain denote flood zone 1 which is regularly flooded in monsoon season (August to September). This zone has the most dynamic geomorphic features such as alluvial islands and mid-channel bars due to sedimentation in the channel belt. Therefore, hamlet density is low in this zone while the vulnerability of hamlets therein is high. Flood zone 2 covers embankment protected part of the active floodplain which

[1] Centre for the Study of Regional Development, School of Social Sciences, Jawaharlal Nehru University, New Delhi, India.
[2] Rural Development Officer, Govt. of Bihar, India.
[3] Centre for Landscape and Climate Research, Department of Geography, University of Leicester, Leicester, UK.
* Corresponding author: rajeshcsrd@gmail.com

is inundated once in 2-3 years due to breaches in embankments. In this zone, hamlet density and flood vulnerability of hamlets are moderate. Flood zone 3 is a part of the older floodplain that was inundated during unprecedented floods of 1998 and 2008. The hamlet density is highest in flood zone 3 due to occasional flooding. Vulnerability of hamlets is low in flood zone 3 compared to that of in flood zone 1 and 2. This study may help planners to protect hamlets from future flood events in the flood prone area of the lower Ghaghara river valley.

KEYWORDS: Flood zone, Mean annual flood, Valley margin, Hamlet density, Channel belt, Older and Active floodplain.

○ INTRODUCTION

Floods and droughts are generally results of the extremities of climate variability that occur throughout the world. The high departure of rainfall, both above and below normal, causes floods and droughts, respectively. Yevyevich (1992) defined floods as "extremely high flows or levels of rivers, whereby water inundates floodplains or terrains outside of the water-confined major river channels. Floods also occur when water levels of lakes, ponds, reservoirs, aquifers and estuaries exceed some critical values and inundate the adjacent land, or when the sea surges on coastal lands much above the average sea level". In relation to the meteorological processes, floods are of different types that are flash, heavy rainfall, coastal and dam failure (Rakhecha and Singh 2009).

India receives ~ 80 per cent of its precipitation in just four months (June to September) during monsoon and Indian rivers also show a high amount of discharge in these monsoonal months (Dhar et al. 1981; Dhar and Nandargi 2003; Mooley and Shukla 1989). Heavy rainfall induced riverine flooding is a recurrent natural disaster in India. Floods are considered as the most devastating phenomena causing loss of lives and public properties and bringing untold misery to the floodplain dwellers in many parts of the country. The fertile and level lands in the vicinity of rivers attract human beings to settle there for agricultural as well as for commercial practices. Thus, the human activities encroaching on flood prone areas are further altering the water chemistry and sediment dynamics of the river (Singh and Awasthi 2010).

According to a report of the National Disaster Management Authority (NDMA), Govt. of India (2008), flood prone areas account for 40 million hectares of land out of the total geographical area (3290 lac hectares) of India. The NDMA report (2008) also highlights that on an average, every year, about 1600 lives are lost due to floods in India. According to a report of working group on flood management and region specific issues for the 12th plan, Planning Commission, Govt. of India, the total flood damages (at the price level of 2010) were about Rs. 8, 12,500 crores from 1953 to 2010. The Planning Commission of India estimated in 2011 that the country is losing approximately Rs. 1805 crores every year due to damage caused to crops, houses and public properties by floods. Especially in the Himalayan river basin of the Indo-Gangatic plain, flood hazard has not only destroyed the lives of humans, cattles and animals but also destroyed crops, infrastructure and derailed the economic activities affecting the socio- economic growth of common people in the rural areas. Due to the lack of proper flood management techniques in Uttar Pradesh, floods in the

Ganga and its tributaries and sub-tributaries i.e. Yamuna, Ghaghara, Rapti and Gomti Rivers damage lives, natural resources and eventually the environment every year. Floods also result in the outbreak of serious epidemics like malaria and cholera in the flood prone areas of these rivers. Despite huge outlays and expenditure on flood management techniques, flood damages to houses and flood affected villages showed an increasing trend during 1973-2012 in flood prone areas of Uttar Pradesh (Fig. 1). According to the 11th plan working group, Uttar Pradesh has the highest flood affected area out of the total flood affected areas in the country. The Ghaghara River brings consistent and severe floods in Faizabad, Gonda, Basti, Ambedkar Nagar, Sant Kabir Nagar, Azamgarh, Gorakhpur, Mau, Deoria and Ballia districts of the state.

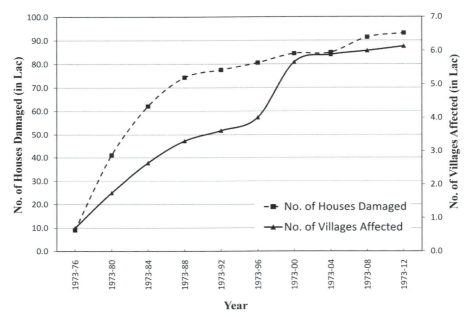

Figure 1 Cumulative flood affected villages and damaged houses in Uttar Pradesh (1973-2012).

Since the introduction of the first five year plan in 1951, both Central and U.P State Govt. have made structural and non-structural measures to control the menace of floods in the Ghaghara river basin but these measures have been proved to be inadequate. Severe floods in Ghaghara River have been causing large scale devastation of public property and loss of lives by submergence of villages, shifting of channel as well as sand casting in the fertile agricultural lands on a regular basis (Planning Commission, Govt. of India, 1981). This can be inferred by a report of the Irrigation Department of Uttar Pradesh according to which about 1516 villages of above mentioned ten districts were affected and about 183 people died during 2008 floods in the lower Ghaghara basin. Thus, it is urgent to adopt a concerted effort to make a critical review of existing flood disaster management by adopting technologically more acceptable and suitable flood hazard management methods in this area.

The flood prone area zonation is a non-structural measure for the assessment and regulation of land use therein to minimize flood induced damages. The main aim of such zonation is to restrict the encroachment of high value land use (buildings, rail and road network) on the active floodplain in order to minimize the flood damages (CWC 1996). In addition to it, the zonation of flood prone area based on geomorphic features, flood magnitude and frequency denotes the assessment of the river valley in terms of susceptibility of floods. Years of research on flood prone area mapping enable earth scientists to use space borne optical satellite images for accurate, quick and temporal mapping of the flooded area (Morrison and Cooley 1973; Tangri 1992; Pandey et al. 2010). However, these days, scientists are also using images of flooded area acquired from all weather satellite sensors (Earth Resource Satellite-1 Synthetic Aperture Radar and IRS-Radar Imaging Satellite). Optical satellite images coupled with ERS-1 synthetic aperture radar (SAR) images and SRTM DEM are often used for real-time estimation of flooded area, flood vulnerability of settlements and 3D flood mapping which are the integral parts of flood hazard management (Wang et al. 1995; Wang 2004; Sanyal and Lu 2004 Sanyal and Lu 2005; Chandran et al. 2006; Jain et al. 2006; Matgen et al. 2007; Ho et al. 2010). The passive optical sensor images such as Landsat multispectral sensor (MSS), thematic mapper (TM), enhanced thematic mapper (ETM+) and IRS LISS-II, LISS-III and advanced wide field Sensor (AWiFS) are utilized for mapping the floodplain geomorphology using the geographic information system (GIS) (Jain and Sinha 2003; Sinha et al. 2008; Gaurav et al. 2011).

Kumar and Singh (1997) attempted to prepare the flood prone zones in lower Rapti-Ghaghara *Doab* on the basis of flood frequency using the traditional cartographic methods. They also discussed channel erosion but did not study the geomorphic features in detail. Haq and Bhuiyan (2004) prepared the high, moderate and low flood magnitude maps using satellite images of RADARSAT-WIFS and *Satellite Pour l'Observation de la Terre* (SPOT) on the 1 : 50 K scale. They used these flood maps along with geomorphology and the maximum water level data to delineate flood zones in Tangail district of Bangladesh. However, they did not address the influence of sediment load and unprecedented floods on channel planform and geomorphology which often cause large scale breaches in embankments. Patel and Srivastava (2013) prepared a flood zone map of Surat city of Gujarat (India) using Google-earth, IRS-1D satellite image, SOI topographical map (1 : 50 K) coupled with the digital elevation model (DEM) based hydrological analysis. The methods discussed in this chapter were selected from the previous research works of the several authors (Graf 2000; Jain and Sinha 2003; Haq and Bhuiyan 2004; Pandey et al. 2010; Sinha et al. 2012). Despite years of research on flood hazard and its zonation, lack of such types of works on the study area prompted us to carry out the present study. Hence, the objective of this chapter is to delineate the flood prone area and divide it into flood zones between Ayodhya and Turtipar by using geomorphological parameters (geomorphic features) and hydrological parameters (magnitude and frequency of floods) and to assess the flood vulnerability of hamlets in each flood zone. In addition to it, this study also discusses the role of sediment dynamics and frequency of floods on altering the channel planform and geomorphology of the lower Ghaghara floodplain.

○ STUDY AREA

The Ghaghara River is a major left bank tributary of the Ganga River. It is a mountain fed river originating from a glacier near Lampia pass at an altitude of 4800 m (amsl). The total drainage area of the basin is ~ 1,207,950 Km² out of which 57,647 Km² area lies in India and the remaining in Nepal and Tibet. The total length of the Ghaghara River is ~ 1080 Km (NIH 1998-99). The study stretch covers ~ 23 per cent (~ 252 Km) of the total length. The annual rainfall ranges between ~ 100 and ~ 150 cm. The Sharda River is a major right bank tributary while Rapti, Kuwano and Chhoti Gandak Rivers are the left bank tributaries of the Ghaghara River (Fig. 2). The stretch of river under study is a part of middle Ganga plain where the thickness of Quaternary alluvial fills varies from ~ 2000 to ~ 3000 m (Singh 1996).

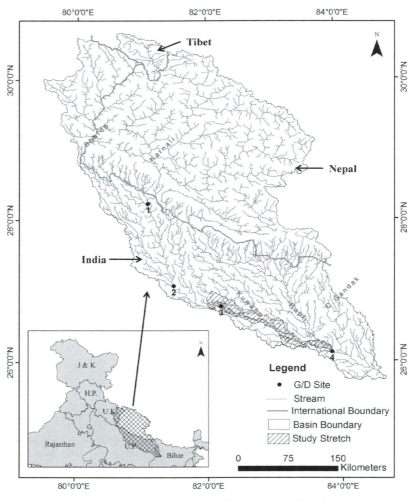

Figure 2 Location map of the study area. The block circle with label shows gauge and discharge sites, Katarniaghat (1), Elgin Bridge (2), Ayodhya (3) and Turtipar (4).

○ DATA BASE

This study is based on space borne medium resolution satellite images and hydrological data. The U.S. Army Map Service (AMS) topographic maps and the flood protection and canal maps of Uttar Pradesh were also used to map the embankments. The Google Earth (GE) and Flood Report of 2008 published by Irrigation Department of Uttar Pradesh were used to identify the spur erosion and breach sites in embankments. Primary channel, channel belt and active floodplain morphology were mapped from the Landsat MSS (1977), TM (1990) and LISS –III (2008) satellite images. The Landsat images were downloaded from USGS earth explorer website (http://earthexplorer.usgs.gov/). Indian Remote Sensing Satellite Resourcesat-1 LISS-III images were downloaded from http://bhuvan.nrsc.gov.in/bhuvan_links.php. Shuttle Radar Topography Mission (SRTM) DEM (ver.4.0) was obtained from http://srtm.csi.cgiar.org/. The extracted flood extent of September 19-20, 1982 that was interpreted from Landsat 3 MSS images by Tangri (1992) was used in this study. Flood extent of 1998, 2005 and 2008 was traced from Dartmouth Flood Atlas Maps (http://www.dartmouth.edu/~floods/Modis.html). Further, the highest flood level (HFL) and the flood days data (1971-2008) recorded at Ayodhya and Turtipar gauge and discharge site were obtained from Irrigation Department of Uttar Pradesh and CWC, New Delhi. A detailed list of data used is given in Table 1.

TABLE 1 Description of Satellite images, topographic sheet, flood atlas and hamlet data sets.

A. Satellite images and DEM data					
Date of acquisition	Satellite	Sensor	Worldwide Reference System (WRS)		Resolution (m)
			Path	Row	
February 14, 1977	Landsat-2	MSS	153	42	60
February 13, 1977	Landsat-2	MSS	152	42	60
September 19, 1982	Landsat-3	MSS	153	041	60
September 20, 1982	Landsat-3	MSS	154	041	60
November 17, 1990	Landsat-5	TM	143	41	30
November 10, 1990	Landsat-5	TM	142	42	30
October 10, November 27, 2008	Resourcesat-1	LISS-III			24
February 11-22, 2000	Shuttle Radar Topographic Mission (SRTM)	SRTM DEM Ver.4 downloaded from http://srtm.csi.cgiar.org/			90

Table 1 Contd.

B. Topographic sheet, Flood atlas and Hamlets data		
Year	Publisher	Scale
1922-25	Army Map Service (RMBM), Corps of Engineers, U.S. Army, Washington, D.C.	1 : 250 K
1998-2009	Dartmouth Flood Atlas, Dartmouth Flood Observatory (http://www.dartmouth.edu/~floods/Modis.html; accessed on 23-12-2009)	Global index map (10 degree display sheet)
1915-1977	Study Stretch Hamlets (Point) data downloaded from http://india.csis.u-tokyo.ac.jp/default/singleHamlet	1 : 50 K & 1 : 63,360

○ METHODOLOGY

Valley margins on the either side of the Ghaghara River were identified using the SRTM DEM, LISS-III, Landsat TM images and maximum flood extent (1998 and 2008). Transect method was used to identify breaks in the elevation profile on either side of the active channel (Sinha et al. 2012) (Fig. 3).

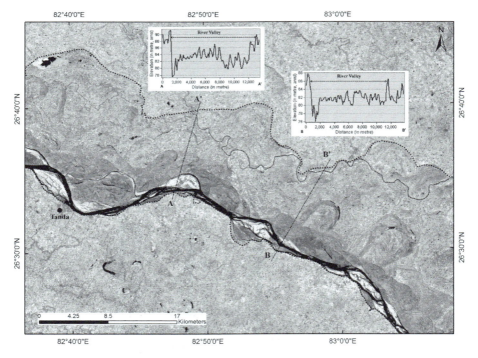

Figure 3 Delineation of river valley using the transect method. The dotted lines either side of river show the river valley margins.

Entire stretch of the river under study is divided into two reaches based on the nature of shift of the river Ghaghara. The general shifting trend of the Ghaghara River along reach-1 is towards south-west while it is towards north-east along reach-2. Further, the morphological characteristics along these reaches were assessed by sinuosity and braided index and geomorphic features within the river valley. Sinuosity (SI) and braiding (BI) indices were calculated using the formula given in equation 1 and 2, respectively (Friend and Sinha 1993).

SI = (Total Central Line Length of the Primary Channel/Valley Length)   (1)
BI = (Total Central Line Length of Secondary segment/Primary Channel)   (2)

where, the primary channel is defined as the widest channel in the stretch under study. The secondary segment or channel shows narrower channel width than the primary channel. The valley length denotes the straight line length between upstream and downstream point of the reach.

Active floodplain was identified using the infrared bands of MSS, TM and LISS-III sensors because this particular band shows the soil moisture content effectively. The geomorphic features of the active floodplain were identified through the visual interpretation of the above mentioned multispectral images. The visual interpretation had taken account of tone, texture, pattern, association, size, shape, topography and temporal behavior of geomorphic features in the active floodplain. On-screen digitization of geomorphic features of the river valley was performed at 30 K scale using ArcGIS-10.

Active channel behavior was assessed using the locational probability model developed by W. L. Graf (2000). The probability model shows the most likely location of the channel over the selected period of time. The active channel mapped from AMS topographic sheet (1922-25) was used as the base channel with a numeric weight (W) of zero. The numeric weights for rest of the channel of the year 1977, 1990 and 2008 were calculated using the equation 3.

$$W = t/m \qquad (3)$$

where, t denotes the separating years between two successive channels while m represents a total time span of the channel.

For instance, the separation time between the channel of 1922-25 and 1977 is ~ 55 years while the total time span for the channel is ~86 years. Thus, the weight for channel of 1977 is ~ 0.64. The most probable location of the channels of different time period is given in Table 2. After assigning the numeric weights for all channels, the union function was applied using ArcGIS-10. The formula for the final locational probability (p-value) is given in equation 4. The final p-values were divided into three classes using the equal interval method that minimizes within group differences and maximizes between group differences as compared to other methods(Tiegs and Pohl 2005).

$$p = (W_{channel\ 1977} + W_{channel\ 1990} + W_{channel\ 2008}) \qquad (4)$$

Flood zones were demarcated using the low, moderate and high magnitude floods occurred in the study area. The flood magnitude was defined on the basis of areal coverage of flood layers and related duration of floods of the year 1982, 1998, 2005 and 2008 (Table 3). The overlay operation between these flood layers and geomorphic units in GIS environment produced three flood zones along the Ghaghara River.

Annual occurrence (P) of floods (in days) at Ayodhya and Turtipar gauge and discharge site was analyzed using the formula given in equation 5 (Raghunath 2006).

$$P = m/y \tag{5}$$

where m denotes number of days when water level remains above the danger level and y represents total period of records.

Tehsil wise hamlet density (no. of hamlets per Km^2) in each zone was calculated in ArcGIS-10 using Kernel point density extension of spatial analyst.

TABLE 2 Locational probability of channel (1922-25 to 2008).

S. No.	Year	Separating period (t, in year)	Weight (W)
1	1922-25	0	0
2	1977	55	0.64
3	1990	13	0.15
4	2008	18	0.21
Total		86	1

TABLE 3 Inundated area of different floods with their magnitude.

Flood year	Inundated area (in Km²)	Duration of floods (in days)		Magnitude
		Ayodhya	Turtipar	
1982	1371	45	23	Moderate
1998	2810	51	70	High
2005	882	22	5	Low
2008	1424	67	60	High

○ RESULTS AND DISCUSSION

Causes and Occurrence of Floods in Lower Ghaghara River Valley

The lower Ghaghara river valley is the most vulnerable to floods because heavy rainfall in the upper catchment area brings huge volumes of water and heavy sediment load which, along with inadequate flood detention capacity in the channel belt, drainage congestion and erosion of river banks, cause severe floods and sand casting in crop lands (NIH 1998-99). Major synoptic systems such as low pressure area over the basin and monsoon trough passing through the basin with a shifting trend towards foothills of Himalaya produce very high rainfall and as a consequence, floods occur in the valley in the month of August and September (Ram and Pangasa 2000). The annual occurrence of floods (in days) at Ayodhya and Turtipar G/D site is ~ 21 and ~ 31 respectively. Sinha et al. (2008) estimated the mean annual flood (2.33 year return period) using the maximum discharges recorded at various sites in Kosi river basin. They also discussed the intensity and variability of flooding (sporadic, regular and extensive) therein on the basis of deviation of the maximum discharge from the mean annual flood. Similarly, the deviation of the highest flood level (HFL) from the mean annual flood level (2.33 year return period) shows the flood intensity and variability at Ayodhya and Turtipar G/D site. Sporadic (irregular) flooding recorded at Turtipar but not at Ayodhya from 1971 to 1981 while the reverse was seen during 1988-90 and 1999-2006. The period 1986-87 and 1991-96 witnessed no major flooding. A regular flooding occurred in the period 1982-84. An extensive flooding was seen in 1998, 2007 and 2008 (Fig. 4). The period 1998 and 2008 witnessed unprecedented floods at Ayodhya and Turtipar G/D site. Changes in the unprecedented flood levels at Ayodhya and Turtipar G/D site are occurring at 10 and 14-15 years intervals, respectively (Table 4).

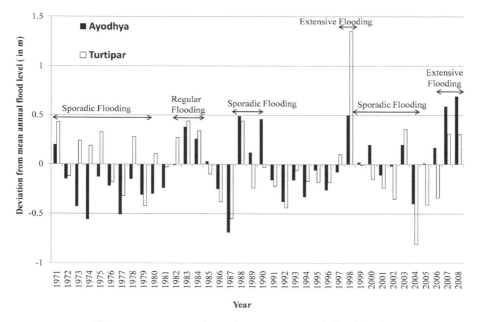

Figure 4 Deviation of HFL from mean annual flood level.

TABLE 4 Unprecedented floods in lower Ghaghara river valley.

G/D Site	DL (m)	HFL (m)	Date	Difference of HFL from DL
Ayodhya	92.73	93.53	1983	0.80
		93.64	August 20, 1988	0.91
		93.65	August 19, 1998	0.92
		93.84	September 25, 2008	1.11
Turtipar	64.01	64.61	September 30, 1969	0.60
		65.09	September 19, 1983	1.08
		66.00	August 28, 1998	1.99

Source: Dhar and Nandagiri 2002 and Irrigation Department of UP.

Geomorphological Features of Flood Prone Area

The lower Ghaghara river valley is divided into two major parts namely older and active floodplain formed by fluvial dynamics. Active floodplain is an area of regular inundation while the older floodplain is occasionally flooded under the current flow regime. However, the major floods (water level remains a meter above the danger level)during 1998 and 2008 caused inundation in older floodplain. For the purpose of this study, the entire stretch is divided into two reaches for a detailed study of geomorphic features therein. The slope along reach-1 and reach-2 is ~ 0.12 m/Km.

Channel belt is a part of active floodplain and contains awe-inspiring geomorphic landforms/features such as alluvial islands, channel, chute channels, abandoned channels, flood deposits, meander cut offs, mid-channel bars, point bars and side bars (Fig. 5). The level of landforms (fluvial origin) is listed in Table 5. In this study, channel belts of the year 1977, 1990 and 2008 have been merged together to identify the channel belt migration pockets. Being a part of former channel, the abandoned channels are partly filled with water and sediments and have no flow (GSI and ISRO 2010). But these channels get reactivated during major floods. Alluvial islands are large sized mid-channel bars with vegetation. Chute channels are elongated channels visible on mid-channel bars and alluvial islands. These channels get activated during flood season. Flood deposits evolve along the cut bank of a meander loop when water spills over the bank during major floods. Meander cutoffs are arcuate or sinuous planform of a meander loop which is not attached to the active channel. These cutoffs have water. Meander scars are sediment filled meander cutoffs and are common features on older floodplain, such scars are also visible along the valley margins (Fig. 5). The horns of these meander scars and cutoffs indicate the direction of the channel shift. Meander scrolls are the ridges of abandoned point bars developed on the slip off slope of meander cutoffs or scars (Charlton 2008). Mid-channel bars denote channel bed accretion surfaces visible in the middle part of active channels during low flow. Point and side bars are laterally accreted surfaces (Ruhe 1975). Point bars are sandy patches without vegetation along the slip off slope of a meander loop while the side bars evolve along the straight part of a channel.

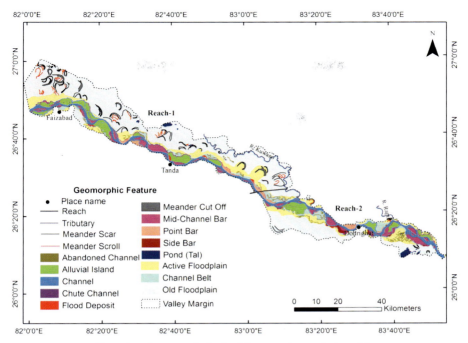

Figure 5 Geomorphological features in reach-1 and 2.

TABLE 5 Classification of geomorphic features of lower Ghaghara river valley.

Level 1	Level 2	Level 3
Valley margin	Older floodplain	Meander cut off
		Meander scroll
		Meander scar
		Abandoned channel
		Tal/lake
	Active floodplain	Channel belt
		Abandoned channel
		Alluvial island
		Channel
		Chute channel
		Flood deposit
		Meander cut off
		Mid-channel bar
		Point bar
		Side bar
		Tal/lake/pond

Channel pattern of the stretch under study shows a low degree of meandering as the sinuosity index values vary between ~1.2 and ~1.3 (Schumm 1980). There is slight straightening of the channel observed during 2008 (Table 6). Therefore, area under alluvial islands is decreasing along reach-1 and 2 due to expansion of the active channel during 1977-2008 (Table 7). This signifies dissection and loss of the vegetal cover of alluvial islands. The side bar area along reach-1 is showing a decreasing trend during 1977-1990 while it shows an increasing trend during 1977-1990 but a decreasing trend along reach-2 during 1990-2008 (Table 8). The point bar area reflects a decreasing trend along reach-1 while this feature is stable along reach-2. Changes in side and point bars reflect shifts in bank lines.

TABLE 6 Sinuosity Index (1977-2008).

1977	Total channel length (Km)	Valley length (Km)	Sinuosity
Reach-1	142	114	1.2
Reach-2	118	91	1.3
Entire stretch	260	205	1.3
1990			
Reach-1	143	114	1.3
Reach-2	116	91	1.3
Entire stretch	259	205	1.3
2008			
Reach-1	138	114	1.2
Reach-2	111	91	1.2
Entire stretch	249	205	1.2

TABLE 7 Areal coverage (Km²) of geomorphic features in reach-1 (1977-2008).

Geomorphic features	1977	1990	2008
Alluvial island	157	162	142
Active channel	65	69	119
Chute channel	34	16	20
Flood deposits	9	5	6
Mid-channel bar	71	79	92
Point bar	46	34	26
Side bar	18	6	2
Channel belt area (CBA)	400	371	407
Total bar area	301	286	268
Bar area/CBA	0.75	0.77	0.66

TABLE 8 Areal coverage (Km²) of geomorphic features in reach-2 (1977-2008).

Geomorphic features	1977	1990	2008
Alluvial island	143	155	75
Active channel	62	62	109
Chute channel	33	14	14
Flood deposits	9	12	3
Mid-channel bar	62	59	90
Point bar	21	22	22
Side bar	8	15	11
Channel belt area (CBA)	339	340	323
Total bar area	244	264	200
Bar area/CBA	0.72	0.78	0.62

Ratio between the total bar and channel belt area reflects the processes of aggradation and degradation along the stretch under study. An aggradation process occurred along both the reaches between 1977 and 1990 while the degradation process was observed during 1990-2008 due to occurrence of high magnitude floods in 1998 and 2008. These processes are largely affected by the occurrence of floods. Annual occurrence of floods at Ayodhya and Turtipar G/D site was ~ 17 and ~ 29 days above the danger level during 1977-1990 while it increased to ~ 28 and ~ 31 days during 1991-2008, respectively.

TABLE 9 Braiding index (1977-2008).

1977	Total secondary channel length (Km)	Primary channel length (Km)	Braided index
Reach-1	290	142	2.0
Reach-2	284	118	2.4
Entire stretch	574	260	2.2
1990			
Reach-1	318	143	2.2
Reach-2	327	116	2.8
Entire stretch	645	259	2.5
2008			
Reach-1	375	138	2.7
Reach-2	303	111	2.7
Entire stretch	678	249	2.7

Average annual and monsoonal silt load data recorded at Elgin Bridge and Turtipar G/D sites are showing the deposition of large amount of sediments in this stretch. The average annual silt load recorded at Elgin Bridge and Turtipar is ~7664.6 and ~7524.1 Th.H.M. (thousand hectare meter), respectively. The average monsoonal silt load at these G/D sites is ~7347.9 and ~7310.2 Th.H.M., respectively (NIH 1998-1999). Variations in average annual and monsoonal silt load between the upstream (Elgin bridge) and downstream (Turtipar) sites are ~140.6 and ~37.7 Th.H.M., respectively. These variations in silt load as well as an increasing trend in unprecedented flood level denote deposition of a large amount of silt load in the midstream area. Therefore, mid-channel bar area shows an increasing trend along both the reaches during 1977-2008. An increasing mid-channel bar area causes primary channel to bifurcate into secondary and chute channels in the channel belt. As a consequence, braiding index values are also showing an increasing trend during 1977-2008 (Table 9).

Flood Prone Area Zonation and Flood Vulnerability of Hamlets

Total river valley area of the Ghaghara River between Ayodhya and Turtipar is ~3954 Km². The total flood prone area (all three zones) covers ~55 per cent area while flood free (older floodplain) zone accounts for ~45 per cent area of the river valley. Sagri, Harraiya, Ghanghata, Madhuban, Basti and Khajni tehsil collectively account for ~76 per cent of the total flood prone area (Table 10 & Fig. 6).

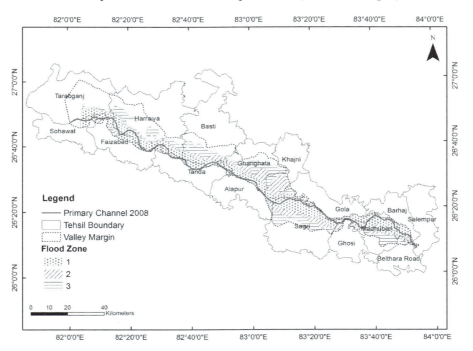

Figure 6 Flood zones of the lower Ghaghara flood prone area.

TABLE 10 Tehsil wise distribution of flood prone area.

District	Tehsil	Area in km²				Percent to total flood prone area			
		Zone 1	Zone 2	Zone 3	Total	Zone 1	Zone 2	Zone 3	Total
Gonda	Tarabganj	56	11	11	78	2.6	0.5	0.5	3.6
Faizabad	Sohawal	46	0	0	47	2.1	0.0	0.0	2.1
	Faizabad	61	18	13	92	2.8	0.8	0.6	4.2
Basti	Harraiya	186	30	122	339	9	1.4	5.6	15
	Basti	77	44	80	201	3.5	2.0	3.7	9
Ambedkar Nagar	Tanda	77	0	0	77	3.5	0.0	0.0	3.5
	Alapur	32	3	9	44	1.5	0.1	0.4	2.0
Sant Kabir Nagar	Ghanghata	78	80	114	272	3.6	3.6	5.2	12
Gorakhpur	Khajni	33	89	61	183	1.5	4.1	2.8	8
	Gola	59	11	8	79	2.7	0.5	0.4	3.6
Azamgarh	Sagri	213	88	128	429	10	4.0	5.8	20
Mau	Ghosi	3	11	4	18	0.1	0.5	0.2	0.8
	Madhuban	138	35	65	239	6	1.6	3.0	11
Deoria	Barhaj	55	10	1	67	2.5	0.5	0.1	3.0
	Salempur	6	0	0	6	0.3	0.0	0.0	0.3
Ballia	Belthara Road	2	5	10	17	0.1	0.2	0.5	0.8
Total		1124	437	627	2188	51	20	29	100.0

Vulnerability is an expected harm or damage to the cultural environment due to floods. It is determined by a group of factors including location and condition of hamlets, infrastructure, socio-economic condition of floodplain dwellers, public policy and administration (Wilhelmi and Wilhite 2002).

Flood zone 1 is regularly flooded in the monsoon season (last week of August to second week of September) when the water level approaches danger level or remains above it by ~0.43 m. Mean annual flood (MAF) level at Ayodhya and Turtipar G/D site is ~93.15 and ~64.65 m, respectively. Since MAF level is higher than the danger level (bankfull level) at these sites, flooding is a regular phenomenon in zone 1. This zone covers ~51 per cent of the total flood prone area. Sagri, Harraiya and Madhuban tehsil in this zone account for ~10, ~9 and ~6 per cent of the total flood prone area, respectively (Table 10). Flood zone 1 mimics the active floodplain and records various episodes of changes in geomorphic features. This zone includes geomorphic features like alluvial islands, channels, chute channels, abandoned channels, flood deposits, meander cut offs, mid-channel bars, point bars and side bars. The mid-channel bar area is increasing and as a consequence, water detention capacity of this zone reduces. Therefore, large beaches occur in recent embankments during floods. The braiding index values are also showing an increasing trend and as a consequence, the channel exerts pressure on the spur of the recent embankments and thus, weakening of embankment occurs. On the basis of the locational probability model, the channel area under low probability is ~65 per cent of the total channel area (Figs. 7a & b). An area under low probability denotes low disturbed area which is mostly occupied by alluvial islands and mid-channel bars. Moderate probability class covers ~18 per cent of the total channel area and shows moderately disturbed area that is generally occupied by the chute channels, alluvial islands and mid-channel bars. The high probability class covers ~17 per cent of area of the total channel area and signifies highly disturbed area that is occupied by the active channel of 2008 (e.g. Tiegs and Pohl 2005). Embankments aligned near to moderately and highly disturbed areas are relatively more susceptible to breach. Older embankments under the channel belt area have been completely destructed by the shifting of active channel (Figs.7a & b). This zone is an area of regular erosion and deposition (Fig. 8). The total number of hamlets in this zone is ~344. The types of houses are generally temporary huts made of thatch and mud (Kumra and Singh 1997). But in some part of this zone, brick and concrete made houses are also located near embankments. The hamlet density varies between ~0.06 and ~0.5 (Fig. 9). This flood zone denotes high vulnerability of hamlets. A high hamlet density is found in Tarabganj, Harraiya and Sagri tehsil (Table 11).

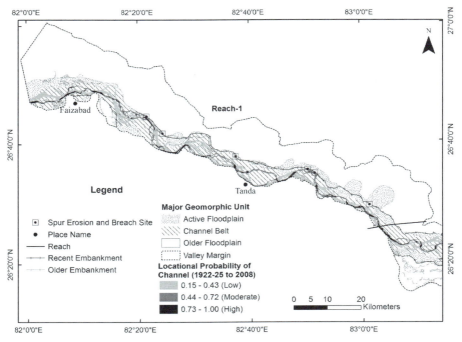

Figure 7a Locational probability of channel in the channel belt of reach-1 (1922-25 to 2008).

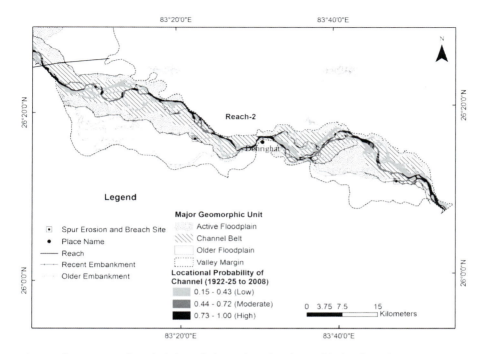

Figure 7b Locational probability of channel in the channel belt of reach-2 (1922-25 to 2008).

Figure 8 The sand casting area of flood zone 1 downstream of Faizabad city (A) and near the Rapti-Ghaghara confluence (B).

Flood zone 2 covers ~20 per cent of the total flood prone area. This zone is inundated during moderate magnitude floods when water level remains between ~0.43 and ~0.91 m above the danger level at Ayodhya (Flood frequency once in ~3 years) and Turtipar G/D site (flood frequency once in ~2 years), respectively. Khajni, Ghanghata, Basti and Harraiya tehsil collectively account for ~50 per cent area of flood zone 2. Breaches in embankments cause flooding and sand casting in this zone. Major geomorphic features of this zone are the active floodplain protected by embankments, abandoned channels and meander cutoffs. The abandoned channels get reactivated during the monsoon season. Total number of hamlets under this zone is ~ 450. Hamlet density varies from ~0.3 to ~2.0 (Fig. 9) and denotes moderate flood induced vulnerability of hamlets. A high hamlet density is observed in Sohawal, Tarabganj, Harraiya, Basti, Ghanghata, Khajni, Gola and Belthara road tehsil (Table 11).

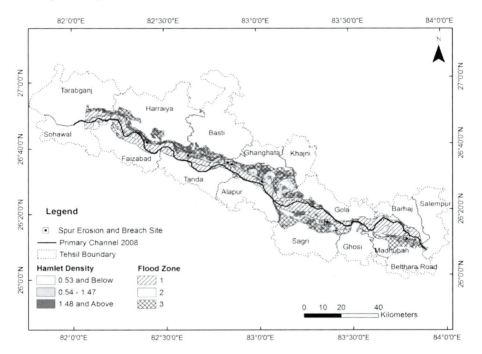

Figure 9 Hamlet density (no. of Hamlets/Km²) in lower Ghaghara flood prone area.

Flood zone 3 accounts for ~29 per cent of the total flood prone area. A large chunk of area in flood zone 3 is distributed among Harraiya, Sagri and Ghanghata tehsil. This zone is occasionally inundated when the water level remains between ~1.0 and ~2.0 meters above the danger level at Ayodhya and Turtipar G/D site, respectively. Major geomorphic features are the older floodplain and its related features like meander scroll bars, meander scars, and the *Tal*(pond). The highest number of hamlets (~784) falls under this zone. The hamlet density varies from ~0.7 to ~2.3 which is slightly higher than that found in flood zone 1 and 2 (Fig. 9) due to occasional flooding. Hence, flood vulnerability of hamlets is low compared to flood zone 1 and 2. Hamlets of flood zone 2 and 3 are made of brick and concrete. A high hamlet density is observed in Alapur, Tarabganj, Harraiya, Basti, Ghanghata, Khajni and Gola tehsil (Table 11).

TABLE 11 Number of hamlets and density in flood prone area.

District	Tehsil	No. of hamlets				Hamlet density		
		Zone 1	Zone 2	Zone 3	Total	Zone 1	Zone 2	Zone 3
Gonda	Tarabganj	31	19	13	63	0.6	1.7	1.2
Faizabad	Sohawal	3	1	0	4	0.1	2.2	0.0
	Faizabad	11	7	9	27	0.2	0.4	0.7
Basti	Harraiya	96	43	190	329	0.5	1.4	1.6
	Basti	23	59	127	209	0.3	1.3	1.6
Ambedkar Nagar	Tanda	7	0	0	7	0.1	0.0	0.0
	Alapur	11.0	1	20	32	0.3	0.3	2.3
Sant Kabir Nagar	Ghanghata	18	80	164	262	0.2	1.0	1.4
Gorakhpur	Khajni	13	87	83	183	0.4	1.0	1.4
	Gola	10	12	11	33	0.2	1.1	1.3
Azamgarh	Sagri	91	87	120	298	0.4	1.0	0.9
Mau	Ghosi	1	8	3	12	0.3	0.8	0.7
	Madhuban	16	31	33	80	0.1	0.9	0.5
Deoria	Barhaj	11	9	1	21	0.2	0.9	0.7
	Salempur	2	0	0	2	0.3	0.0	0.0
Ballia	Belthara Road	0	6	10	16	0.0	1.1	1.0
Total		344	450	784	1578			

○ CONCLUSIONS

The analysis aptly demonstrated the detailed mapping of flood prone area coupled with distribution of hamlets in each of the flood zones. Each flood zone shows the distinctive geomorphic features and flood characteristics. Flood zone 1 covers the most dynamic geomorphic features such as channel belt and bars. When flood level approaches danger level or remains above it by 0.43 m, this zone gets inundated. Sedimentation in zone 1 has caused rise in bed level as a consequence, the level of unprecedented floods is increasing. An increasing trend in the mid-channel bar area and braiding index values also signifies sedimentation in the channel belt. Such sedimentation in channel belt of flood zone 1 triggers the processes of breaches in embankments and as a consequence, flood zone 2 experiences flooding once in ~2-3 years. Afforestation in mountainous and foot hills areas should be encouraged to reduce the silt load and to retard the frequency of unprecedented floods in the lower Ghaghara river valley. Flood zone 3 is occasionally gets flooded when flood level remains above danger level by ~1 to ~2 m. Hamlet density is increasing from flood zone 1 to flood zone 3 while the flood vulnerability of hamlets is decreasing from zone 1 to zone 3 due to variations in flood magnitude and frequency. In terms of the flood prone area, high hazardous tehsils are Harraiya, Basti, Ghanghata, Khajni, Sagri and Madhuban. Unprecedented flood level is increasing in the flood prone area due to siltation in river bed therefore the flood induced damages to hamlets can be minimized in each flood zone by adopting the flood proofing measures (raising the elevation of the hamlets) that is a highly appreciated flood mitigation measure in Uttar Pradesh. The data on hydrology, temporal changes in channel and floodplain geomorphology given in this study will be beneficial for policy makers and earth scientists in controlling the adverse impacts of floods. It will be also beneficial for floodplain dwellers if these data would be added in current programs on flood control measures to protect embankments and to perform flood proofing for reducing the flood vulnerability of settlements located in the flood prone area.

○ REFERENCES

Central Water Commission (CWC). 1996. Flood Plain Zoning. Pamphlet No. 1/96, New Delhi.

Chandran, R.V., Ramakrishnan, D. and Chowdhary, V.M. 2006. Flood mapping and analysis using air-borne synthetic aperture radar: a case study of july 2004 flood in Baghmati river basin, Bihar. Current Science 90: 249-256.

Charlton, R. 2008. Fundamentals of Fluvial Geomorphology. Routledge, London.

Dhar, O.N., Rakhecha, P.R. and Mandal, B.N. 1981. Some facts about Indian rainfall – a brief appraisal from hydrological considerations. Indian Journal of Power & River Valley Development 31(7&8): 117-125.

Dhar, O.N. and Nandargi, S. 2003. Flood study of the Himalayan tributaries of the Ganga river. Meteorological Applications 9: 63-68.

Friend, P.F. and Sinha, R. 1993. Braiding and meandering parameters. In: Best, J.L. and Bristow, C.S. (eds.). Braided Rivers, Special Publication 75. Geological Society of London. 75: 105-111.

Gaurav, K., Sinha, R. and Panda, P.K. 2011. The Indus flood of 2010 in Pakistan: a perspective analysis using remote sensing data. Natural Hazard 59: 1815-1826.

Graf, W.L. 2000. Locational probability for a dammed, urbanizing Stream: Salt River, Arizona, USA. Environmental Management 25(3): 321-335.

GSI & ISRO. 2010. Manual for national geomorphological and lineament mapping on 1:50000 scale. Hyderabad, 1-112.

Haq, Kazi Md. F. and Bhuiyan, R.H. 2004. Delineation and zonation of flood prone area: a case study of Tangail district, Bangladesh. Indian Journal of Regional Sciences 36(1): 20-30.

Ho, L.T.K., Umitsu, M. and Yamaguchi, Y. 2010. Flood hazard mapping by satellite images and SRTM dem in the Vu Gia-Thu Bon alluvial plain, central Vietnam. The International Archives of the Photogrammetry, Remote Sensing and Spatial Information Sciences, XXXVIII, 275-279.

Irrigation Department of Uttar Pradesh. 2008. Flood report-2008, Research and Planning (Flood) Division, Lucknow, 1-23.

Jain, S.K., Saraf, A.K., Goswami, A. and Ahmad, T. 2006. Flood inundation mapping using NOAA AVHRR data. Water Resource Management 20(6): 949-959.

Jain, V. and Sinha, R. 2003. Geomorphological manifestations of the flood hazard: a remote sensing based approach. Geocarto International 18(4): 51-60.

Kumra, V.K. and Singh, J. 1997. Flood prone zoning of lower Rapti-Saryu doab: a geographical approach. 46-58. *In*: Nag, P., Kumra, V.K. and Singh, J. (eds.). Geography and Environment, Vol. II. Concept Publishing Company, New Delhi.

Matgen, P., Schumann, G. and Henry, J.B. 2007. Integration of SAR-derived river inundation areas, high-precision topographic data and a river flow model toward near real-time flood management. International Journal of Applied Earth Observation and Geoinformation 9: 247-263.

Mooley, D.A. and Shukla, J. 1989. Main features of the westward moving low pressure system which forms over the Indian region during the summer monsoon season and their relation to the monsoon rainfall. Mausam 40: 137-152.

Morrison, R.B. and Cooley, M.E. 1973. Assessment of flood damage in Arizona by means of ERTS-1 imagery. pp. 755-760. *In*: Proceedings of Symposium on Significant Result Obtained from the Earth Resource Satellite 1, Vol. 1. New Carrollton, Maryland.

National Institute of Hydrology (NIH), 1998-99. Hydrological inventory of river basins in eastern Uttar Pradesh. Jal Vigyan Bhawan, Roorkee, SR-1/98-99, 20-53.

NDMA. 2008. National Disaster Management Guidelines: Management of Floods. Govt. of India, 19.

Pandey, A.C., Singh, S.K. and Nathawat, M.S. 2010. Waterlogging and flood hazards vulnerability and risk assessment in Indo-Gangetic plain. Natural Hazards 55: 273-289.

Patel, D.P. and Srivastava, P.K. 2013. Flood hazards mitigation analysis using remote sensing and GIS: correspondence with town planning scheme. Water Resource Management 27: 2353-2368.

Planning Commission. 1981. Report on development of chronically flood affected areas. National Committee on the Development of Backward Areas. Govt. of India, New Delhi.

Planning Commission. 2011. Report of Working Group on Food Management and Region Specific Issues for XII plan. Govt. of India, New Delhi, 18.

Raghunath, H.M. 2006. Hydrology. New Age International (P) Limited, New Delhi.

Rakhecha, P. and Singh, V.P. 2009. Applied hydrometeorology. Concept Publishing Company, New Delhi.

Ram, L.C. and Pangasa, N.K. 2000. Semi quantitative precipitation forecasts for Ghaghara catchment by synoptic analogue method. Mausam 51(1): 85-100.

Ruhe, R.V. 1975. Geomorphology: Geomorphic Processes and Surficial Geology. Houghton Mifflin Company, Boston.

Sanyal, J. and Lu, X.X. 2004. Application of remote sensing in flood management with special reference to monsoon Asia: a review. Natural Hazards 33: 283-301.

Sanyal, J. and Lu, X.X. 2005. Remote sensing and GIS-based flood vulnerability assessment of human settlements: a case study of Gangetic West Bengal, India. Hydrological Processes, 19: 3699-3716.

Schumm, S.A. 1980. Planform of alluvial rivers. Proceedings of the International Workshop on Fluvial River Problems held at Roorkee, India, 4-21 to 4-4.

Singh, D.N. and Awasthi, A. 2010. Natural hazards in Ghaghara river area, Ganga plain, India. Natural Hazards 57: 213-225.

Singh, I.B. 1996. Geological evolution of Ganga plain-an overview. Journal of the Palaeontological Society of India 41: 99-137.

Sinha, R., Bapalu, G.V., Singh, L.K. and Rath, B. 2008. Flood risk analysis in the Kosi river basin, north Bihar using multi-parametric approach of analytical hierarchy process (AHP). Journal of the Indian Society of Remote Sensing 36: 293-307.

Sinha, R., Jain, V., Tandon, S.K. and Chakraborty, T. 2012. Large river systems of India. Proc Indian Nat Science Academy 78(3): 277-293.

Tangri, A.K. 1992. Satellite remote sensing as a tool in deciphering the fluvial dynamics and applied aspects of Ganga plain. pp. 73-84. *In*: Singh, I.B. (ed.). Gangetic Plain: Terra Incognita. Geology Department, Lucknow University.

Tiegs, S.D. and Pohl, M. 2005. Planform channel dynamics of the lower Colorado river: 1976-2000. Geomorphology 69(1-4): 14-27.

Wang, Y., Koopmans, B.N. and Pohl, C. 1995. The 1995 flood in the Netherlands monitored from space-a multi-sensor approach. International Journal of Remote Sensing 16: 2735-2739.

Wang, Y. 2004. Using Landsat 7 TM data acquired days after a flood event to delineate the maximum flood extent on a coastal floodplain. International Journal of Remote Sensing 25: 959-974.

Wilhelmi, O.V. and Wilhite, D.A. 2002. Assessing vulnerability to agricultural drought: a Nebraska case study. Natural Hazards 25: 37-58.

Yevyevich, V. 1992. Floods and society. pp. 3-9. *In*: Ross, G.H.N. and Yevyevich, V. (eds.). Coping with Floods. NATOASI Series, Dordrecht, Kluwer.

7

CHAPTER

◇◇◇

Geospatial Technology for Water Resource Development in WGKKC2 Watershed

Vandana Tomar,[1] *Anand N. Khobragade,*[2] *Indal K. Ramteke,*[3]
Prem Chandra Pandey[4,]* and *Pavan Kumar*[5]

ABSTRACT

Water is the natural resource which is facing the problems like scarcity, overexploitation etc. It needs the immediate actions to be conserved and managed. Watershed provides a natural unit to provide the better management and development of an area in terms of executing development plans. Watershed management leads to development of the area along with uplifting the standard of living for all beings. In the present study, the Water Resource Development Plan (WRDP) has been prepared using the merged ortho-product of LISS-IV and Cartosat-1D by considering the various criteria. The watershed approach depends on the various factors of area i.e. land capability, soil depth, landuse/landcover, slope, drainage and soil erosion which are taken in to consideration for the present study. By creating and following the decision rules for different conservation structures, the WGKKC2 watershed was partitioned in Continuous Contour Trenching (CCT), Earthen Nala Bunds (ENB), Cement Nala Bunds (CNB), Loose Boulder Structure (LBS) and Gully Plugs (GP). The weighted overlay analysis and spatial query has been used for zonation of the watershed, in which the different themes have been

[1] Research Officer, Haryana Institute of Public Administration (HIPA), Gurgaon, India.
[2] Resource Scientist, Maharashtra Remote Sensing Applications Center, Nagpur, India.
[3] Scientific Associate, Maharashtra Remote Sensing Applications Center, Nagpur, India.
[4] Centre for Landscape and Climate Research, Department of Geography, University of Leicester, Leicester, UK.
[5] Department of Remote Sensing, Banasthali University, Newai, Tonk, Rajasthan, India.
* Correspondence author: prem26bit@gmail.com;pcp6@alumni.le.ac.uk

assigned a particular weightage by considering their contribution to reach the goal. The conservation structure zoning provides the better use of natural resources and conserving both soil and water resources.

KEYWORDS: WGKKC2 Watershed, LISS-IV, Sustainable Development.

○ INTRODUCTION

The watershed approach is increasingly being deployed in various development programs to manage the water and land resources like soil and water conservation, dry land or rainfed farming, ravine reclamation, control of shifting cultivation, etc. Management of land and water resources in rainfed areas by integrating remote sensing and GIS is required for optimum development of land and water resources and to meet the basic minimum needs of people in a sustained manner. One of the major gaps in the watershed development program has been the inadequate database for planning and to conduct research on the methodology and implementation (Dhruvanarayana et al. 1990; Vidyanathan 1991; Sarkar and Singh 1993). The watershed approach depends on the various factors of the area i.e. geology, geomorphology, soil, slope, run-off etc (Kumar, 2010). The watershed is a dynamic and unique place. Watershed approach has been the single most important landmark in the direction of bringing in visible benefits in rural areas (Shankaret al. 2012; Marble et al. 1983; Gugan et al. 1993; Tideman et al. 2000). The watershed provides the overall production of an area such that from resources management to employment generation. It also improves the rural livelihood by uplifting of the rural people economically and socially i.e. the development leads to education, employment and agricultural development.

Management of a watershed entails the rational utilization of land and water resources for optimum production while causing minimum trauma to natural and human resources. The watershed forms an appropriate unit for analyzing the development-linked resource problems, designing the appropriate solutions of identified problems and eventually testing the efficacy of the prescribed solutions (Anon 1982; Tejwani 1987). Watershed management tries to bring about the best possible balance in the environment between natural resources on the one side and human and other living beings on the other (Vora, 2011). Watershed management is an iterative process of integrated decision making regarding uses and modification of lands and waters within a watershed. Development of the watershed needs better understanding about the various natural resources their relations with each other and their relations with livelihood of the stakeholders. Kushwaha et al. (1996) have demonstrated the method for integrated sustainable rural development planning using remotely sensed data and GIS. The water resource development plans includes proposal of water conservation measures for deep percolation, increase of water table, decrease the run-off potential for soil conservation, fulfilling the requirement of water bodies and water structures, intensifying the agricultural as well as non-agricultural land. Water conservation relates to soil and water conservation in the watershed, which includes: rainwater conservation, land Management and run-off conservation.

Remote Sensing (RS) has become an indispensable scientific tool for mapping and monitoring of natural resources and frequently used for planning, characterization and management of natural resources. GIS and various satellite Sensors are being used for various levels and scales in watershed area for mapping, monitoring, planning, execution and implementation. The integration of Remote Sensing and GIS techniques has emerged as a powerful tool for watershed-based conservation measures by suitably identifying sites for conservation structures (Saptarshi and Raghvendra 2009). Combination of Analytical Hierarchy Process and Compromise Programming techniques worked well in solving Single Objective Multi-Criteria problems like Site Selection for Water Harvesting Structure, Landslide Hazard Zonation (Novaline et al., 2001; Prasada et al., 2001; Prasada et al., 2002). In this study, weighted overlay analysis using remote sensing and GIS has been applied to identify the natural resources problems and to generate a locale specific watershed development plan. The integration of Remote Sensing and GIS techniques has emerged as a powerful tool for micro-watershed based conservation measures by suitably identifying sites for conservation structures (Durbude and Venkatesh 2004). Integration of the two technologies has proven to be an efficient tool in groundwater studies (Krishnamurthy et al. 1996). By interfacing remote sensing with GIS, different management scenarios could be generated, which help the planners toassess the feasibility of various alternatives before selecting the one that would be most suitable (Neillis et al. 1990). RS data and GIS tools academics along with topographical maps and ground truth should be enhanced to control watershed deterioration and implement management strategies. Effective management tools can be designed only with backup from extensive RS data and various GIS tools. Identification and delineation of Land Use/Land Cover (LULC), location, extent and their spatial distribution patterns are possible through RS because of its synoptic view and its ability to resolve both macro and micro details on satellite imagery.

Extensive ground truth with GPS (Global Positioning System) provide better and reliable results which can be compared with digital images. GPS plays a vital role in collecting the actual location of soil and water conservation structures in the watershed. We can show these structures on maps, so that planners can make plans for conservation of the soil and water easily and promptly.

○ STUDY AREA

WGKKC-2 watershed lies in Nagpur District and covers three tehsil of Nagpur which are Katol, Kalmeshwar and Nagpur rural. It lies on latitude 21°5'8.12"N and longitude 79°2'48.52"E covering an area of 360 km². The location map of study area has been shown in Fig. 1. The area has a tropical wet and dry climate with dry conditions prevailing for most of the year. It lies on the Deccan plateau of the Indian Peninsula and has a mean altitude of 310.5 meters above sea level. The region experiences tropical climate and records the rise of temperature up to 48°C in summer season (March to May). The cold season is from December to February and temperature drops down to as low as 6°C to 8°C. The southwest monsoon season is from June to September while the period October-November constitutes the post-monsoon season. The average annual rainfall in Nagpur District is 1161.54 mm.

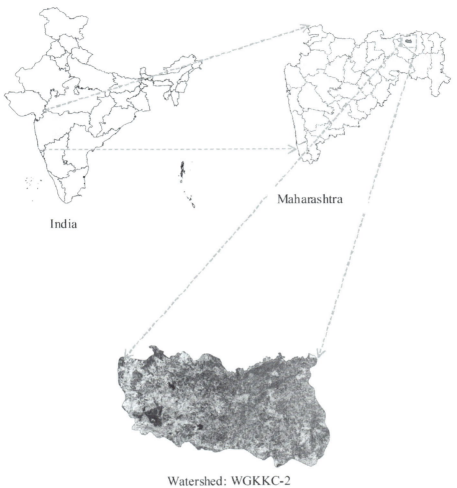

India

Maharashtra

Watershed: WGKKC-2
District: Nagpur

Figure 1 Location map of study area.

○ MATERIAL AND METHODOLOGY ══════════

Data Used

High resolution dataset has been used for generation of the thematic layer i.e. CARTOSAT-1D, LISS-IV Multi-spectral (MX) to prepare the water resources development plan upto the village level for rural development. A merged product of orthorectified CARTOSAT-1D and LISS-IV MX has benn taken in the present study. SOI toposheet is very helpful for preparing the base map and geocoding the satellite imageries in the case of digital image interpretation. GPS points of soil and water construction structures, location of wells (with photo), location of some important features such as field plantation, wasteland in watershed were acquired during extensive field work, so that these structures could be located and marked on the satellite image accurately. These points will help to cross verify the results in which the proposed conservation structures and already established structures can be accurately overlaid and matched.

Image Processing

The data sets are prepared with toposheet at $1:50000$ scale. Water resource development plans are addressed for proposing the conservation structures. This may lead to the improvements in the present condition or suggest strengthening the activities. In this we have used the various themes i.e. slope, land capability, soil depth, land use/land cover and soil erosion. This integration has been done in GIS environment and weighted overlay analysis is being used for the proposal of the WRDP. The methodology chart used in the present study is illustrated in Fig. 2.

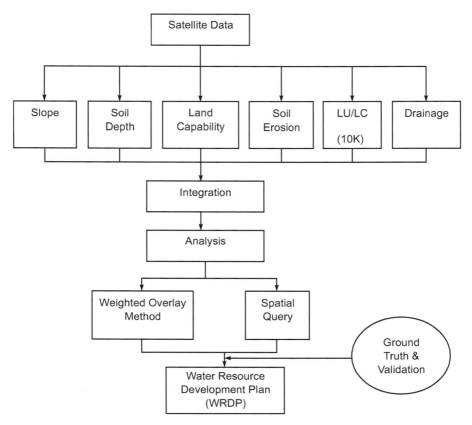

Figure 2 Methodology Chart of the study.

○ RESULTS AND DISCUSSION

Land Use/Land Cover is demarcated with the help of SOI toposheet and 50 K data. The crop land covers about 72% of the total area, whereas the wasteland covers about 14% of total land area. The forest, including both open and dense areas, covers about 7% area of the total land. The water bodies have an area of about 2%. The transportation covers area about 0.22%. The built up area covers area of about 3% whereas industrial area contributes to about 1% of the total area. LULC map has been shown in Fig. 3.

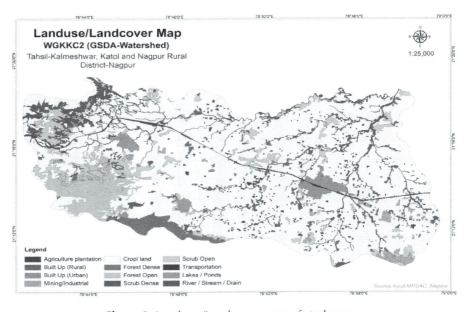

Figure 3 Land use/Land cover map of study area.

Based on the information available about the area and its related parameters, we can suggest different water and soil conservation measures. The information availability of the area with the consortium of this project on research findings on the suitability of different land use management technologies for given agro climatic conditions, individual combinations of various resource themes, have to be recommended for specific activities. The water resources development and management influence the land resource management to a great deal. Therefore it is necessary to develop, conserve and efficiently utilize the available water in the soil profile. The present approach adopted for suggesting alternative sustainable land use comprises taking into consideration present land use/land cover, soils and slope. Water resource plan is prepared by following the guide lines enumerating suitable site conditions for structures for surface water harvesting, groundwater recharge and exploitation and efficient methods for irrigation.

The decision rules for various water conservations structures are shown in Table 2. The CNB can be made at second or third order streams where flow is maximum, erosion should be slight. The slope should be less and soil should be very deep. The LBS and ENB have the same criterion i.e. soil should be deep and erosion should be slight to moderate, the land capability should be second class and slope not more than 5-10% but the difference is that ENB can be constructed over where earth is available and LBS can be constructed in forest areas where the boulders are available. The GP can be proposed where erosion can be moderate to severe and soil depth can be shallow or moderately deep and slope can be from 1-5%. These should be proposed in scrub land. The CCT should be proposed at higher slope and severe erosion i.e. at hilly area where land is totally non arable and soil is very shallow. The decision making is flexible with the analytical approach which includes classification.

TABLE 1 Decision rules for WRDP.

Cons. type	Criteria				
	Land capability	Erosion	Slope%	Depth	Land use/ Land cover
CCT	V, VI	Severe	15-35	Very shallow	Not suitable for agriculture
GP	III, IV	Moderate to severe	1-3/3-5	Shallow/ moderately	Open scrub
LBS	III	Slight to moderate	5-10	Deep	Forest
ENB	II	Slight to moderate	5-10	Deep	Agriculture
CNB	I	Slight	0-5	Very deep	Agriculture

Each theme has a different contribution towards the goal. In weighted overlay method, the layers are assigned with weights according to the contribution of the different themes for the central theme. The slope is given as higher weightage as it contributes the maximum to get the information about the type of water harvesting structure. It reveals many facts about the physiography of the land. It has been given the weightage of 0.4 at the scale of 0 to 1. Then follow soil erosion and land capability, which are given as same weightage i.e. 0.2. The soil erosion shows the amount of eroded soil in the land so that water flow can be minimized and erosion of soil could be stopped. The soil depth and LU/LC is given the weightage of 0.1. The soil depth reveals the fact of magnitude of vegetation at that place which in turn indicates the soil capability to hold the water flow. The classes of themes have been given rank according to the relevance in theme. Like slope classes are ranked orderly between the scales of 1 to 5, similarly the soil erosion's order is between 1 to 6 and land capability as 1 to 4. The other themes are soil depth which has the scale of 1 to 4 and land use/land cover which is ranked as 1 and 2 as shown in Table 2.

TABLE 2 Weightage and ranking of criteria for WRDP.

S. No.	Layers	Classes	Rank	Weight
1.	Slope	1-3%	1	0.4
		3-5%	2	
		5-10%	3	
		10-15%	4	
		15-35%	5	
2.	Soil erosion	None-slight	1	0.2
		Slight	2	
		Slight-moderate	3	
		Moderate	4	
		Moderate-severe	5	
		Severe	6	
3.	Land capability	IIes	1	0.2
		IIew	1	
		IIIes	2	
		IIIs	2	
		IVe	3	
		VIes	4	
4.	Soil depth	Very deep (> 100 cm)	1	0.1
		Deep (50 to 100 cm)	2	
		Moderately deep (25 to 50) cm)	3	
		Shallow (10 to 25 cm)	4	
5.	Land use/Land cover	Agriculture plantation	1	0.1
		Scrub dense	1	
		Scrub open	1	
		Cropland	2	

The WRDP zonation map (Fig. 4) is proposed using different methods to choose the best applicable criterion and best method suited for development of WRDP. The area statistics of zonation of WRDP has been shown in Table 3 and Figure 5.

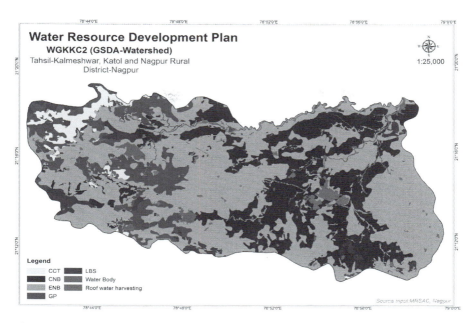

Figure 4 Water Resource Development Plan at WKKC2 watershed Nagpur, Maharashtra India.

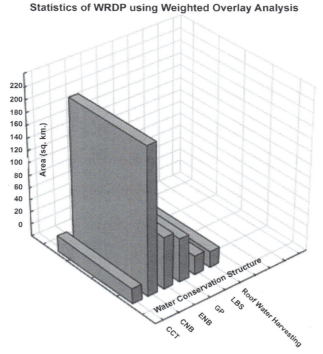

Figure 5 Statistics of WRDP using weighted overlay analysis.

TABLE 3 Area statistics of WRDP using weighted overlay method.

Water conservation structure	Area (sq. km.)	Percentage
CCT	5.33	1.52
CNB	215.8	61.56
ENB	63.57	18.13
GP	49.77	14.20
LBS	10.14	2.89
Roof water harvesting	5.93	1.69
Total	350.54	100.00

○ CONCLUSIONS

Watershed developmental activities provide soil and water conservation measures, reducing scarcity and agricultural development. The watershed development includes the overall improvement of an area alongwith socio economic development, employment, education. Various themes have been considered to propose the conservation structures such as Land use/Land cover, drainage, slope, soil depth, land capability and soil erosion. The study on thematic mapping revealed that utilization of remote sensing techniques and ArcGIS in conjunction with traditional methods have been found a very useful methodology to prepare various thematic maps and tabular information in order to generate sustainable development activities for water resources. For the purpose of water conservation related to WRDP, the weighted overlay and spatial query techniques have been used for proposal of water harvesting structures by CNB and ENB is provided in an agricultural area by considering the higher order streams in the area, with minimum slope and erosin and maximum soil depth. Loose boulder structure, gully plugging and continuous contour trenching were suggested in the regions with steeper slopes to reduce the runoff rate and soil erosion, in forest and scrub areas with shallow soil depth. Rain water harvesting technique is suggested for the rural and urban settlement areas for collecting and storing the rain water from the building roof for domestic use as well as to improve ground water.

○ REFERENCES

Anon. 1982. Socio-economic development of Himalayan Hills, Task force for the study of eco-development of Himalayan region, Planning Commission, Government of India, New Delhi.

Dhruvanarayana, V.V., Shastry, G. and Patnaik, U.S. 1990. Watershed Management. ICAR, Krishi Anusandhan Bhavan, Pusa, New Delhi.

Durbude, D.G. and Venkatesh, B. 2004. Site suitability analysis for soil and water conservation structures. Journal of Indian Society of Remote Sensing 32(4): 399-405.

Gugan, D.J. 1993. Integration of remote sensing and GIS. Proceedings of the International Symposium on the Operationalization of Remote Sensing. Enschede, ITC, pp. 77-85.

Krishnamurthy, J.N., Venkatesa, K., Jayaraman, V. and Manivel, M. 1996. An approach to demarcate ground water potential zones through remote sensing and geographical information system. International Journal of Remote Sensing 17: 1867-1884.

Kumar, D. 2010. Project Report, Application of GIS in Watershed Management. pp 1-39. Available online http://www.aiggpa.mp.gov.in/files/pdf/intern%20scheme/RGM_dheeraj.pdf

Kushwaha, S.P.S., Subramanian, S.K., Chennaiah, G. Ch., Ramana Murthy, J., Rao, S.V.C., Perumal, A. and Behera, G. 1996. Interfacing remote sensing and GIS methods for sustainable rural development. International Journal of Remote Sensing 17: 3055-3069.

Marble, D.F., Peuquet, D.J., Boyle, A.R., Bryant, N., Calkin, H.W. and Johnson, T. 1983. Geographic information system and remote sensing. pp. 923-958. *In*: Colwell, R.N. (ed.). Manual of Remote Sensing. American Society of Photogrammetry, Falls Church, VA.

Nellis, M.D., Lulla, K. and Jensen, J. 1990. Interfacing geographic information system and remote sensing for rural land use analysis. Photogrammetric Engineering and Remote-Sensing 56: 329-331.

Novaline, Jacob, Saibaba, J., Prasada, Raju, Krishnan, R. and Subramanian, S.K. 2001. Site selection for water harvesting structure in watershed development using Decision Space. Proceedings of Geomatics – 2001 conference, Published by Indian Institute of Remote Sensing, Dehradun, India, 45-49.

Prasada, R., Novaline, J. and Saibaba, J. 2001. Decision Space – A tool for modelling landslide hazards. Proceedings of the International Conference on Remote Sensing & GIS/GPS (ICORG-2001), Published by JNTU, Hyderabad, India.

Prasada, R., Novaline, J., Saibaba, J., Rao, A.S., Das, B.K. and Sujatha, G. 2002. Decision Space software for locating suitable water harvesting structures in parts of Alwar district, Rajasthan. Proceedings of the National conference on GIS application in micro level planning, Published by National Institute for Rural Development, Hyderabad, India.

Saptarshi Praveen, G. and Raghvendra Rao, Kumar. 2009. GIS-based evaluation of micro-watersheds to ascertain site suitability for water conservation structures. Journal of Indian Society of Remote Sensing 37: 693-704.

Sarkar, T.K. and Singh, A. 1993. Interactive approach for watershed management – A case for interaction. Tech. Rep. No.WTC/IARI/WSM. IARI, New Delhi, pp. 3-93.

Shankar, P., Hermon, R.R., Alaguraja, P. and Manivel, M. 2012. Comprehensive water resources development planning Panoli village blocks in Ahmednagar district, Maharashtra using Remote Sensing and GIS techniques. International Journal of Advances in Remote Sensing and GIS Vol. 1, No. 1.

Tejwani, K.G. 1987. Watershed management in the Indian Himalaya. pp. 203-227. *In*: Khoshoo, T.N. (ed.). Perspectives in Environmental Management. Oxford & IBH, New Delhi.

Tideman, E.M. 2000. Watershed Management: Guidelines for Indian Conditions. Omega Scientific, New Delhi.

Vaidyanathan, A. 1991. Integrated Watershed Development (supplement): Some Major Issues. Foundation day Society for Promotion of Wastelands Development, New Delhi 6(4): 1-19.

Vora, Krunali. 2011. Application of Remote Sensing & GIS in Watershed land use development. National Conference on Recent Trends in Engineering & Technology (B.V.M. Engineering College, V.V.Nagar, Gujarat, India), 13-14 May.

Section III

Satellite Based Approaches

CHAPTER

Predicting Flood-vulnerability of Areas Using Satellite Remote-sensing Images in Kumamoto City, Japan

A. Besse Rimba* and *Fusanori Miura*

ABSTRACT

Flood is a natural disaster that occurs almost every year in Japan. According to the flood record, it occurs during the rainy season, around July of each year. The aim of this paper is to predict areas vulnerable to flooding in the Shiragawa watershed. This study was carried out using DEM data, ALOS AVNIR-2 imagery and Amedas data to produce a watershed area, a vegetation index, a land-cover map and an isohyet map. DEM data, with spatial resolution of 10 meters, were derived from the Geospatial Information Authority of Japan (GSI) to show topography and the watershed. The ALOS AVNIR-2 image was used to create the land-cover map and the vegetation index. The land-cover map was created by the unsupervised method and then verified by use of the land-cover map of the Geospatial Information Authority of Japan (GSI). The vegetation index was created by use of the Normalized Difference Vegetation Index (NDVI) algorithm. The isohyet map was obtained by use of data from rain-gauge stations in Kumamoto Prefecture and then interpolating by the application of the kriging method. All spatial data were overlaid to create the flood-vulnerability map by application of the Geographic Information System (GIS) model. This study combines all the data to predict areas vulnerable to flood. The result indicates that flood occurs in the middle part of the Shiragawa watershed.

KEYWORDS: Flood, satellite imagery, Geospatial Information Authority of Japan, Geographic Information System, Shiragawa watershed.

Graduate School of Environmental Science and Engineering, Yamaguchi University, Tokiwadai, Ube, Yamaguchi, Japan.
* Corresponding author: a.besserimba@yahoo.com

115

○ INTRODUCTION

Flood is one of the most devastating natural disasters that regularly occur in Asia, even in developed countries such as Japan. Flood is caused not only by rainfall but also by such events as ice-jam clogs, failure of riverbanks, storm surges, tsunamis, high tides, rainstorms, lake outbreaks and slope failures (Excimap 2007).

Flood is influenced by many factors, such as the intensity and duration of storms, topography, geologic structure, vegetation cover and hydrologic conditions (Konrad 2000). On July 12, 2012, flood in Kumamoto Prefecture on Kyusu Island was caused by heavy rain on Mt. Aso, where the water level exceeded 2,300 m^3/s (Manu et al. 2014). The torrential rainfall caused significant landslides in the area around Mt. Aso and the largest area affected by the flood was the Shiragawa River, which has flat topography in the middle and downstream. Rainfall lasted more than 24 hours, with an accumulation of 507.5 mm.

The impact of a flood can be measured by the number of victims and the amount of economic damage (Guha-Sapir et al. 2013). According to a report by the Japan Society of Civil Engineering Western Branch in Kyushu on the flood in Kumamoto Prefecture, 30 people died, 2 people were missing, 11 people were injured, 313 houses were completely destroyed and 1,500 houses were partially destroyed. Hence, this paper predicts the flood-vulnerable area to prevent loss of life and damage to property in the future by applying the remote sensing technology and the Geographic Information System (GIS) model.

Remote sensing technology has an important role in this paper. Advanced Land Observing Satellite (ALOS) Advanced Visible and Near Infrared Radiometer-2 (AVNIR-2) imagery, digital elevations models (DEMs) and rainfall data were used to produce flood parameters. The topography analysis was derived by the application of DEMs. The vegetation index, rainfall distribution and land-cover data improved the reliability of the spatial analysis as well as that of the flood-vulnerability area. The high resolution of DEM images gives the topography information in detail. The vegetation index and land-cover data give information about the land-surface cover. The rainfall data show the distribution of precipitation, it is one parameter that involved the flood-vulnerability area. Combining some parameters can be done by use of the Geographic Information System (GIS) model because each parameter in the GIS model has a score. The flood-vulnerability area can be predicted by overlaying and scoring all parameters.

○ STUDY AREA AND METODOLOGY

Research Location

The length of the Shiragawa River is approximately 78 km, with the headwaters at Mt. Aso and the downstream in Kumamoto City. The topography of the Shiragawa River is mountainous near the headwaters and flat in the middle and downstream. Because of this topography, even when no rain falls in the middle and downstream parts of the river, flood is possible from rain on Mt. Aso. The water flows from Mt. Aso to Kumamoto City in approximately 2 hours.

According to the flood history of the study area, Kumamoto City has experienced floods on August 30, 1980, July 2, 1990 and July 12, 2012, with the water level reaching an estimated 2,300 m³/s in each flood.

Geographic Information Systems (GIS) and Remote Sensing

The Geographic Information System (GIS) is a computer-based system that handles georeferenced data: (1) data capture and preparation, (2) data management, including storage and maintenance, (3) data manipulation and analysis and (4) data presentation (Otto 2009).

Map is the most known representative in the GIS environment that small-scale symbols some part of the actual phenomena in the earth (Otto 2009). Modelling is the process of producing an abstraction of the real world so that some part of it can be more easily handled. Modelling might be achieved through direct observations by use of sensors and digitizing (converting) the sensor output for computer use (Otto 2009). By detecting the energy that is reflected from the Earth, remote sensors collect data. These sensors can be on satellites or mounted on aircraft. Remote sensors can be either passive or active. Passive sensors respond to external stimuli. The ALOS AVNIR-2 sensors are passive sensors. They record radiation, usually from the sun, that is reflected from the Earth's surface. Because of this feature, passive sensors can only be used to collect data during daylight hours. In contrast, active sensors use internal stimuli to collect data about the Earth. For example, a laser-beam remote-sensing system projects a laser onto the surface of the Earth and measures the time required for the laser to reflect back to the sensor (NOAA 2015).

Data Processing

We used three types of satellite data in this research: ALOS AVNIR-2 imagery to derive the vegetation index and land cover, a DEM from the Geospatial Information Authority of Japan (GSI) and rainfall data derived from Amedas and downloaded from the Japan Meteorology Agency (JMA) website. The DEM was derived to get slope data and the river pattern. Rainfall data were rainfall averages in the rainy season (July 2012) in Kumamoto Prefecture. The reason for using the GSI Map and Google Earth in this research is verification of the parameters of this research. In this case, the land cover and the Normalized Difference Vegetation Index (NDVI) are verified by these data.

ALOS AVNIR-2

The Radiometric Distortion Correction: Pixel values represent the radiance of the surface in the type of digital numbers (DNs), which are calibrated to adequate a convinced range of values. Sometimes the DNs are denoted as the brightness values. Transformation of DN values to absolute radiance values is an essential process for qualified analysis of several images acquired by diverse sensors. Because each sensor has its own calibration parameters used in noting the DN values, the same DN values in two images recorded by two different sensors may denote two different radiance values (Chander et al. 2009).

Vegetation Index: The NDVI is a calculation that impress into the amount of infrared radiation reflected by vegetation. Live green vegetation engages solar radiation, which they use as a source of energy for photosynthesis. The NDVI is frequently used to observe drought, to observe and forecast agricultural production, to aid in forecasting hazardous fire areas and to represent desert encroachment. The NDVI is desired for global vegetation observing because it can recompense for changing enlightenment conditions, surface slope, aspect and other extraneous factors (Lillesand 2004). The idea that the NDVI is useful for sensing vegetation is that fit vegetation reflects excellently in the near-infrared band of the electromagnetic spectrum. Green leaves have a reflectance of 20% or less in the 0.5- to 0.7-micron range (green to red) and approximately 60% in the 0.7- to 1.3-micron range (near-infrared). These spectral reflections are ratios of the reflected over the received radiation in all spectral bands individually; hence, they take on values between 0.0 and 1.0. Thus, the NDVI itself varies between −1.0 and +1.0 that expressed as follows Equation 1 (ArcGIS 2013).

$$NDVI = \left(\frac{IR - R}{IR + R} \right) \tag{1}$$

where:

IR = Pixel values from the infrared band
R = Pixel values from the red band

Land Cover: Land-cover assessment is one of the highly important parameters inland-resources management. Land use control (LUC) inventories are supposedly increasing prominence in various resource sectors, such as agricultural planning, settlement surveys, environmental analyses and operational planning based on agro-climatic zones. Spatial land-cover information is necessary for proper management, planning and monitoring of natural resources (Zhu 1997). Satellite remote-sensing imagery is a viable source for assembly quality land-cover information at the local, regional and global scales (Csaplovics 1998) as is shown in Table 1.

TABLE 1 Land-use/land-cover classification scheme by land-use cover types stratified according to the U.S. Geological Survey's land use and land cover classification system for use with remote sensor data (Anderson et al. 1976).

Item	Description
Build-up land	Area that is used for residential, commercial, industrial, transportation and other facilities.
Forest of rangeland	Area dominated by mature trees, shrubby plants and other plants that grow up with high density.
Water	Area covered by water, such as rivers and lakes.
Agricultural land	Area that are used as rain-fed cropping, planted and irrigated cropping areas.
Barren land	Area with no vegetation cover, degraded land, all unused areas and hilly/mountainous areas.

Digital Elevation Model: A DEM contains terrain elevations for ground location at frequently spread out horizontal intervals. DEMs can be expended for the generation of three-dimensional visuals that display topography slope, aspect and topography profiles amongst selected points. DEMs have developed an important source of geographical data for many scientific and engineering uses, for example, hydrological and geological analyses, infrastructure planning and environmental uses. If topographical data are unobtainable, DEMs from remotely sensed data can be used instead. DEM generation method and DEM quality measurement apply in disaster studies, such as pre- or post-earthquake events and volcanoes (e.g., d'Ozouville et al. 2008).

The Isohyet: An Isohyet is an imaginary line on a weather map that connects points which have equal amounts of precipitation during one period of time. An isohyet map is a map that displays precipitation data. The contoured lines connect areas of equal rainfall and a colour scale is often used to differentiate between areas. Isohyet maps are prepared by interpolating rainfall data recorded at gauged points (VenTe Chow 1964).

The method to draw the isohyet map: (1) arithmetic mean method; the simplest method by which average rainfall is recorded at a number of rain gauges. It is reasonable if the rain gauges are homogeneously distributed over the study area and the individual gage measurements do not diverge, (2) relative weight, if some rain gauges are considered more representative of the study area than others, (3) the theissen method: all points in the watershed catchment are the same as that at the nearest rain gauge, thus the depth recorded at a given gauge is applied out to a distance halfway to the next station in any direction. The relative weight for each gauge is determined from the corresponding areas of application in a Thiessen polygon network. The boundaries of the polygons are formed by the perpendicular bisectors of lines joining adjacent gauges. The theissen method is more accurate than the arithmetic-mean method, but it is in flexible because a new Thiessen network must be constructed each time a change is made in the gauge network, such as when data are missing from one of the gauges. Moreover, the thiessen method does not directly account from orographic influences on rainfall. Some of these difficulties are overcome by the construction of isohyets using observed depths at rain gauges and interpolation between adjacent gauges.

Other methods of weighting rain-gauge records have been offered, such as the reciprocal distance squared method in which the effect of rainfall at a gauged point on the computation of rainfall at an un-gauged point is contrariwise proportional to the distance between the two points (Wei and McGuinness 1973). Singh and Chowdhury (1986) studied the several methods for computing areal average precipitation, including those explained above and concluded that all the methods give a comparable result, especially when the time period is long.

Flood Parameters

TABLE 2 Watershed characteristic as flood/run-off parameters (Ven T. Chow, with modification).

Parameters	Units	Category	Score	Data source	Description
Slope	30%	steep	20	Topography map DEM	Topography map GSIdem
	10-30%	mountainous	15		
	5-10%	surge	10		
	0-5%	relatively flat	5		
Vegetation cover	Vegetation high density	low	5	Satellite interpretation	Satellite imagery
	Vegetation medium density	moderate	10		
	Vegetation low density	high	15		
	Settlement hardened surface	high	20		
Rainfall	<20 mm/24 h	low	5	• AMEDAS	• Satellite imagery
	21-50 mm/h	moderate	10		
	51-100 mm/h	high	15		
	>100 mm/h	extreme	20		

The Equal-interval Classification (constant class intervals): The total score from Table 2 can be classified by use of the equal-interval classification. In this classification method, each class consists of an equal-data interval along the dispersion graph shown in the equation. To determine the class interval total score, divide the range of all data (highest data value minus lowest data value) by the number of classes.

$$\text{Class interval} = \frac{\text{range of data}}{\text{number of classes}} \qquad (2)$$

After determining the class interval, add the result to the lowest value of the dataset, which gives the first class interval. Then add this interval as many times as necessary to reveal the number of predefined classes. The values of the classifications are shown below:

Class 1 (very low)	: 15 – 24
Class 2 (low)	: 25 – 33
Class 3 (medium)	: 34 – 42
Class 4 (high)	: 43 – 51
Class 5 (very high)	: 52 – 60

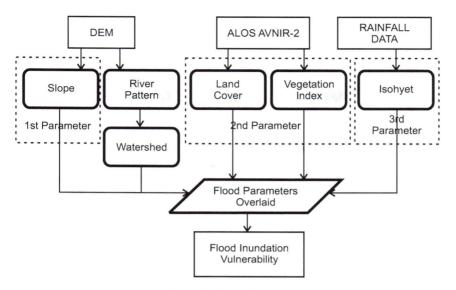

Figure 1 Flow diagram.

○ RESULT AND DISCUSSION

Shiragawa Watershed Map

The Shiragawa watershed map was derived from DEM data with a spatial resolution of 10 meters. DEM point was converted to raster by use of the point-to-raster tool in the ArcGIS 10 software to determine DEM data and delineated the base of the valley and the ridge around the Shiragawa River. The coverage areas of the Shiragawa watershed are Aso City, Minamiaso Mura, Ozu Machi, Kikuyo Machi, Koshi City and Kumamoto City. According to the map as shown in Fig. 2, the total area of the Shiragawa watershed is 559.678606 km².

Slope Map

The slope map was derived from GSIDEM data by use of ArcGIS 10 and then converted from slope in percentage to slope in degree. The slope of the Shiragawa watershed is a hilly step-to-mountainous area in Minamiaso Mura. Aso City, Kikuyo, Koshi City and Kumamoto City are relatively flat, but some areas in eastern part of Kumamoto City are surge areas. According to the slope map (Fig. 3), the Shiragawa watershed area is a relatively flat and mountainous area. In Aso City, which is located on Mt. Aso, the topography is flat even in this highland area. This area is an accumulation of pyroclastic material, which causes the Shiragawa watershed to be rich in alluvium. The area of slope in the Shiragawa watershed can be seen in Table 3. It shows the degrees of slope, the area with a slope of 0%–5% has a high distribution in the Shiragawa watershed.

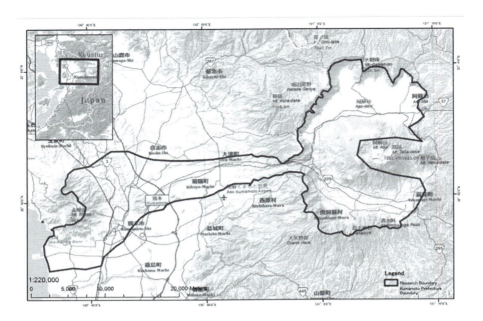

Figure 2 The Shiragawa watershed and research boundary map.

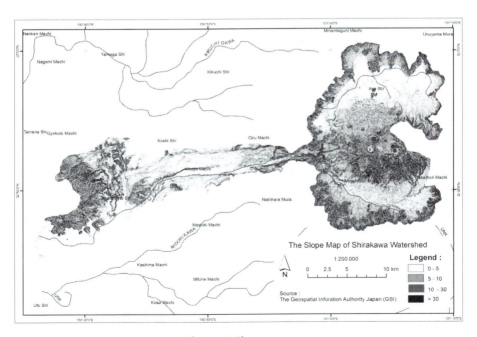

Figure 3 Slope map.

TABLE 3 Total area of the slope of the Shiragawa watershed.

Slope	Area (m²)	Area (km²)
>30%	388,138	0.388138
10%-30%	55,732,112	55.732112
5%-10%	91,914,136	91.914136
0%-5%	411,644,220	411.64422
Total	559,678,606	559.678606

Fig. 3 shows that the west side of the Shiragawa watershed is a mountainous area and on the western side is Mt. Aso, which is the headwater area. In the middle of the Shiragawa watershed is a flat area and in the east is a surge and relatively flat area. This pattern makes the middle part of Shiragawa watershed hollow and then it becomes a pool during the rainfall season.

Vegetation Index

The NDVI was derived from ALOS AVNIR-2 imagery by use of a vegetation index. The NDVI was used in this study to determine the land-cover area. The land cover and the NDVI were correlated to determine the flood-vulnerability parameter. The result shows that the area of the Shiragawa watershed has high vegetation density in Minamiaso Mura, which is an upper area. It has a range value between 0.3-0.58, which indicates that this area has the highest vegetation density. The middle area of the Shiragawa watershed has a low value, which indicates this area has low vegetation or a hard surface. The distribution of the vegetation index can be seen in Fig. 4. It shows the total area of vegetation in each category of the vegetation index.

TABLE 4 Total area of the vegetation index in the Shiragawa watershed.

Vegetation index	Area (m²)	Area (km²)
(−0.75-0.1)	213,397,776	213.3978
(0.1-0.3)	110,917,970	110.918
(0.3-0.58)	235,362,860	235.3629
Total	559,678,606	559.6786

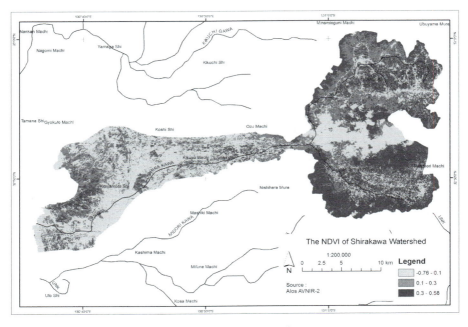

Figure 4 Vegetation index.

Land-cover Map

The land-cover map was derived from ALOS AVNIR-2 imagery. The unsupervised classification was applied to determine this map. This map was verified by use of the Google map and the land cover of the Geospatial Information Authority of Japan (GSI). It was divided into six classes: cloud, water bodies, mixed farm, forest or high vegetation, hard surface and bare land. The ALOS AVNIR-2 imagery was used to determine the land cover, but we could not get information on the areas covered by clouds. The area of land cover for the Shiragawa watershed can be seen in Table 5.

TABLE 5 Total area of land cover in the Shiragawa watershed.

Land cover	Area (m²)	Area (km²)
forest	23,5362,860	235.36286
hard surface	50,240,836	50.240836
bare land	80,557,840	80.55784
mixed farm	110,917,970	110.91797
water body	42,492,680	42.49268
cloud	40,106,420	40.10642
Total	559,678,606	559.678606

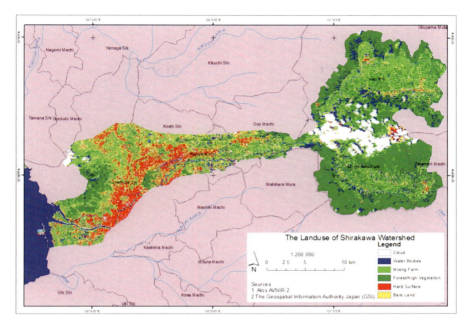

Figure 5 Land cover map.

Rainfall Map

The result shows that the highest rainfall distribution in July 2012 was on Mt. Aso at Aso City. The material of this area is pyroclastic from Mt. Aso, which contains alluvium. If high rainfall occurs in the headwater of the Shiragawa watershed, the middle or the downstream of the watershed will be affected. The rainfall in the middle and lower part of the Shiragawa watershed are 20-30 mm/day. These amounts are not high enough to cause a flood if the headwater area of the watershed has a low rainfall amount also. According to this map, the dominant value is 20-30 mm/day. In the case of the Shiragawa watershed, a flood can occur in the middle or downstream of the watershed if the upstream rain is heavy. The total area of rainfall distribution can be seen in Table 6.

TABLE 6 Total area of rainfall distribution in the Shiragawa watershed.

Rain	m²	km²
<20 mm/day	13,900,684	13.900684
21-30 mm/day	487,602,050	487.60205
41-50 mm/day	58,175,872	58.175872
Total	559,678,606	559.678606

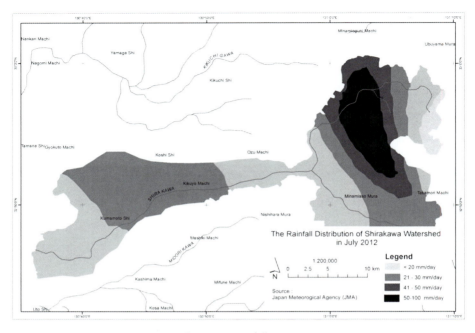

Figure 6 Rainfall map.

Flood-vulnerability Map

The flood-vulnerability map is the final map for this research. It was created by combining three parameters: the slope map, the land-cover map and the rainfall map. The weighting method for determining the class value, in which each feature on the map has a value, was used. This value was determined by use of the Ven T. Chow method, with modification and has equal weight. The level of vulnerability was categorized into five classes: very low, low, medium, high and very high. The classes were determined by use of the equal-interval method, which is the maximum range value subtracted by the minimum range value and then divided by number of classes. By applying Equation 2, the interval class of classification is 8. By use of this method and the pattern of parameters, the highest vulnerable area is determined to be in the middle of Shiragawa watershed. The cloud is still covering the information because this research used only image data as the primary data and all the image data were processed from satellite data. The total area of the flood-vulnerable area can be seen in Table 7. The distribution of flood-vulnerable areas can be seen in Fig. 7.

The vulnerability map shows that the lower area has high flood vulnerability. For example, the rainfall becomes surface runoff in the settlement areas because of the hard road surfaces and less vegetation density. The headwater Shiragawa watershed has low flood vulnerability, because this area has high vegetation density and it was covered by forest. On the other hand, this area has a steep and mountainous slope. Therefore, slope failure can occur in this area.

TABLE 7 Total area of flood vulnerability in the Shiragawa watershed.

Vulnerability	m²	km²
Very low	17,465,699	17.4657
Low	242,119,156	242.1192
Medium	184,561,400	184.5614
High	114,324,860	114.3249
Very high	1,207,491	1.207491
Total	559,678,606	559.6786

To determine the flood-vulnerability area, these parameters should be of use. In this case, vegetation cover should protect this watershed. However, this area is always flooded, even though it is covered by vegetation. The most vulnerable area is approximately 115.59 km², out of the total area of 559 km². This research is only one way to produce the vulnerability map. The vulnerability area in this paper was produced on the basis of satellite imagery only. This result needed to be verified by field survey and historical flood data.

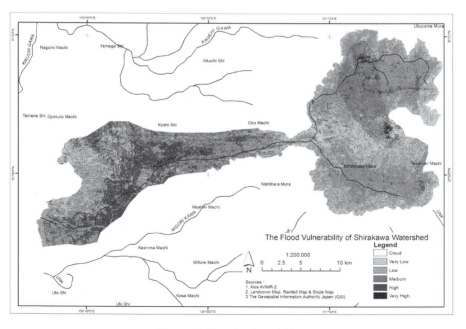

Figure 7 Flood vulnerability.

○ **CONCLUSION**

This research produced the flood-vulnerability area by use of ALOS AVNIR-2 imagery, DEMs and the rainfall data from the AMEDAS data. These methods are some that can be used to determine the flood-vulnerability area. On the basis of this research, the vulnerability area is located in Kumamoto City because this area has a high score in land-cover parameter, high score in slope parameter and middle score in rainfall parameter. This area is covered by hardened surface and has a relatively flat slope. The high rainfall in the headwater of the Shiragawa watershed caused flooding in the middle part of the watershed.

In the headwater of the Shiragawa watershed, the flood did not occur because this area has mountainous topography and a high density of land-cover vegetation. Even though this area has high rainfall distribution, other parameters were not suitable for causing floods. Floods may not occur, but other natural disasters, such as landslides, can occur.

○ ACKNOWLEDGMENT

The author acknowledges JAXA (Japan Aerospace Exploration Agency) and the Ministry of Education and Culture Indonesia, which awarded the author a "Beasiswa Unggulan" Scholarship to support this research. JAXA and Geospatial Information Authority of Japan (GSI) provided data. I also thank Yamaguchi University and Udayana University.

○ REFERENCES

Anderson, J., Hardy, E., Roach, J. and Witmer, R. 1976. A land use and land cover classification system for use with remote sensor data. Geological Survey. Washington: United States Government Printing Office.

ArcGIS, "NDVI function." Accessed 2013, May 25. http://resources.arcgis.com/en/help/main/10.1/index.html#/NDVI_function/009t00000052000000/

Chander, G., Markham, B.L. and Helder, D.L. 2009. Summary of current radiometric calibration coefficients for Landsat MSS, TM, ETM+ and EO-1 ALI sensors. Remote Sensing of Environment 113: 893-903. doi:10.1016/j.rse.2009.01.007

Chow, Ven Te, Maidment, David, R. and Mays Larry, W. 1964. Applied Hydrology. California: McGraw-Hill International Edition Civil Engineering Series. Accessed 2013, June 18. http://www.scribd.com/doc/29283580/Applied-Hydrology-by-Ven-Te-Chow-David-R-maidment-Larry-W

Csaplovics, E. 1998. High resolution space imagery for regional environmental monitoring—status quo and future trends. International Archives of Photogrammetry and Remote Sensing 32(7): 211-216.

D'Ozouville, N., Deffontaines, B., Benveniste, J., Wegmuller, U., Violette, S. and de Marsily, G. 2008. DEM generation using ASAR (ENVISAT) for addressing the lack of freshwater ecosystems management, Santa Cruz Island, Galapagos. Remote Sensing of Environment 112(11): 4131-4147. doi:10.1016/j.rse.2008.02.017

EXCIMAP (European Exchange Circle on Flood Mapping). 2007. Handbook on good: practices for flood mapping in Europe. Endorsed by Water Directors 29-30 November 2007. [Cited 2013, May 15]. Available from: http://ec.europa.eu/environment/water/flood_risk/flood_atlas/pdf/handbook_goodpractice.pdf

Guha-Sapir, D., Hayois, P. and Below, R. 2012. Annual Disaster Statistical Review 2012: The number and trends. Centre for Research on the Epidemiology of Disasters (CRED), Université catholique de Louvain – Brussels, Belgium.

Japan Aerospace Exploration Agency, "ALOS Data Users Handbook Revision C. 2008. Accessed 2013, April 17.
http://www.eorc.jaxa.jp/ALOS/en/doc/fdata/ALOS_HB_RevC_EN.pdf

Konrad, C.P. 2000. Effects of Urban Development on Floods. Geological Survey Water- Fact Sheet 076-03, U.S. Accessed 2013, June 07. http://pubs.usgs.gov/fs/fs07603/.

Lillesand, T., Kiefer, R. and Chipman, J. 2007. Remote Sensing and Image Interpretation—Sixth Edition. John Wiley & Sons, Inc., the United State of America.

Manu, L., Tsukamoto, T., Nakanishi, K., Shirozu, H., Hokamura, T., Nakajo, S., Kuriyama, Y. and Yamada, F. 2014. Long-term evaluation of Shiragawa River Delta due to the extreme Events. Journal of Japan Society Civil Engineers, Ser. B3 (Ocean Engineering) 70: 2.

NOAA. Accessed 2014, February 25th) http://noaa.gov/

Otto, Huisman and Rolf A. 2009. An Introductory Textbook—Principles of Information Systems. ITC, Enchede.

Singh, V.P. and Chowdhury, P.K. 1986. Comparing methods of estimating mean areal rainfall. Water Resource Bulletin 22(2): 275-282. Doi: 10.1111/j.1752-1688.1986.tb01884.x

Wei, T.C. and McGuinness, J.L. 1973. Reciprocal distance squared methods. A Computer Technique for Estimating Area Precipitation ARS-NC-8, U.S. Agric. Research Service of North Central Region, Coshocton, OH. 013, May 25.
https://archive.org/details/reciprocaldistan08weit

Zhu, A.X. 1997. Measuring uncertainty in class assignment for natural resource maps under fuzzy logic. Photogrammetric Engineering and Remote Sensing 63(10): 1195-1202.

Validation of Hourly GSMaP and Ground Base Estimates of Precipitation for Flood Monitoring in Kumamoto, Japan

Martiwi Diah Setiawati,[1,*] *Fusanori Miura*[1] and *Putu Aryastana*[2]

ABSTRACT

GSMaP (Global Satellite Mapping Precipitation) satellite rainfall estimates are evaluated at the hourly time scales and a spatial resolution 0.1° latitude × 0.1° longitude. The reference data came from AMEDAS (Automated Meteorological Data Acquisition System) station network of about 27 rain gauges over Kumamoto Prefecture, Japan. This region has very complex terrain and humidity which is characterized by typhoon. Hence, this area are often hit by flash flood. The research has been conducted to evaluate hourly GSMaP (i.e., GSMaP_MVK (Moving Vector with Kalman Filter) and GSMaP_NRT (Near Real Time) data with AMEDAS data during flood events from 2003 to 2012 and to define the rainfall pattern which causes flood. Statistical analysis was used to evaluate the GSMaP data, both qualitative and quantitative. The results indicate that GSMaP_MVK was reasonably good at detecting precipitation events and GSMaP_NRT was inadequate to represent the rainfall AMEDAS data. Long term and short term rainfall patterns were observed over Kumamoto Prefecture before the occurrence of the flood.

KEYWORDS: GSMaP, verification, flood, Kumamoto.

[1] Graduate School of Environmental Science and Engineering, Yamaguchi University, Tokiwadai, Ube, Yamaguchi, Japan.
[2] Civil Engineering Department, Warmadewa University, Denpasar, Indonesia.
* Corresponding author: s503wf@yamaguchi-u.ac.jp, martiwi1802@gmail.com

⊙ INTRODUCTION ═══════════════════════

Rainfall amount and its spatial distribution are important for flood prediction and water resources assessment (Shrestha et al. 2011). Japan and other countries have been greatly damaged by floods in the past due to heavy rainfall. In addition, Japan is particulary vulnerable to flooding due to its steep topography and humidity characterized by typhoons (Kazama et al. 2009). Moreover, the number of floods in Japan have increased since 2004, especially in the Kyusu region (Kazama et al. 2009). Therefore, a flood forecasting system using rainfall data observed by satellite would be a welcome development. Recently, several kinds of global precipitation satellite data have become available. Some of them have resolutions of one hour and one degree, which may be defined as high temporal and spatial resolution. The GSMaP (Global Satellite Mapping Precipitation) data, as the highest temporal and spatial resolution satellite data, can detect a precipitation event with the same trend as rain gauge data, but the precipitation amount generally has been underestimated (Fukami 2010; Kubota et al. 2009; Makino 2012; Seto et al. 2009; Shrestha et al. 2011). Underestimation of precipitation can cause underestimation of discharge and it causees high bias for flood forecasting (Kabold and Suselj 2005; Pauwels and Lannoy 2005). Hence evaluation of this product is necessary.

Other researchers have shown that GSMaP data products have been verified well in monthly and daily rain gauge data. Shrestha et al. (2011) found that GSMAP_MVK+ performed better in flatter terrain than in the high mountain area over the Central Himalayas. In addition, Kubota et al. (2009) showed that rainfall estimates of GSMaP were the best over the ocean and were the worst over mountainous regions. Seto et al. (2009) noted that monthly GSMaP data had been verified well in Japan, so GSMaP data seemed to be good enough for flood detection.

GSMaP_MVK was verified from January through December 2004 in Japan to determine whether monthly data, daily data and 3 hourly data matched rain gauge data. The result showed that GSMaP_MVK of monthly, daily and 3 hourly data from May to October had 0.7, 0.7 and 0.6 of correlation coefficient and had the same trend as rain gauge data (Kubota et al. 2009). Although monthly, daily and 3 hourly data have been verified, hourly GSMaP data have not yet been verified especially in Kyushu, Japan. Hourly rainfall data is important to understand the rainfall pattern, especially when extreme rainfall occurs. The aims of this research were to verify hourly GSMaP data with rain gauge data and to define the rainfall pattern which causes flood.

⊙ METHOD ═══════════════════════

Study Area

Kumamoto Prefecture is located in the west central Kyushu Island, Japan. The study area covers an area of 389.53 km² from latitude 32°5′45″N to 33°6′17″N and longitude 129°59′8.75″E to 131°19′7.7″E. Kumamoto has a humid subtropical climate and has an elevation ranging from 2 mto 1193 m above the sea level. Precipitation occurs throughout the year with the heaviest in the summer season, especially in the months of June and July. In the summer season from 1981 to 2010, the variability of temperature range was from 12.23°C to 37.5°C and the average of precipitation was 326.4 mm/month.

Data Sets

GSMaP Data

GSMaP was initiated by the Japan Science and Technology Agency (JST) in 2002 and has been promoted by the Japan Aerospace Exploration Agency (JAXA) Precipitation Measuring Mission (PMM) science team since 2007 to produce a global precipitation product with high temporal and spatial resolution (Ushio et al. 2009). The GSMaP product is the combination from low orbit multi satellite microwave radiometer data, such as Tropical Rainfall Measuring Mission Microwave Imager (TRMM TMI), AQUA Advance Microwave Scanning Radiometer (AMSRE), Advance Earth Observing Satellite (ADEOS) II AMSRE and Defense Meteorological Satellite Program Special Sensor Microwave Imager (DMSP SSM/I) and Geosyncronous Orbit (GEO) infra red radiometer data (Okamoto et al. 2007). Brightness temperature at microwave frequencies as the input of GSMaP system was converted into precipitation data. The algorithm to regain surface precipitation rate based on the Aonashi et al. 1996 study was conducted. The combination technique to produce 0.1 degree/1 hour resolution with the domain covering 60° N to 60° S was obtained using a morphing technique using an infra red cloud moving vector and Kalman Filter technique (Ushio et al. 2009). This product was called GSMaP_MVK. GSMaP_MVK version 5 were used in this study.

GSMaP_NRT (near real time) is one of GSMaP products which uses the same algorithm as GSMaP_MVK and after four hours of observation, data can be obtained (EORC and JAXA 2013a). GSMaP_MVK version 5 data are available from March 2000 until December 2010 while GSMaP_NRT are available from October 2008 until now. Hourly data of GSMaP_MVK and GSMaP_NRT for two weeks before the flood occurrence in the past 10 years in Kumamoto Prefecture were downloaded. Both GSMaP_MVK and GSMaP_NRT were processed by using Open GRADS software. Point by point analysis was conducted to compare between rain gauge data and satellite data. This method was also applied by As-syakur et al. 2011.

Rain Gauge Data

Hourly observed rainfall data in Kumamoto Prefecture was obtained from AMEDAS (Automated Meteorological Data Acquisition System) data which was developed by the Japan Meteorological Agency (JMA). There are 36 rain gauge stations available in Kumamoto Prefecture, but only 27 rain gauge stations were used. These data are available online on the JMA website (http://www.data.jma.go.jp) and the distribution of the rain gauge stations is shown in Fig. 1. Rain gauge data which represent the rainfall at the point were used as reference in our comparison. The point-by-point analysis conducted to compare rain gauge data and satellite data is shown in Fig. 2. As-syakur et al. (2011) and Prasetia et al. (2013) also applied this method. Opengrads software was used to adjust the rain gauge data and satellite data. Table 1 summarizes the major specifications of rainfall data for GSMaP product and rain gauge data. We analyzed both data products which have the same temporal resolution that is 1 hour. The GSMaP product domain is 60°N-60°S, but in this research only Kumamoto Prefecture was analyzed. AMEDAS data are available from November 1974 until now and the separation distance of each rain gauge is approximately 21 km.

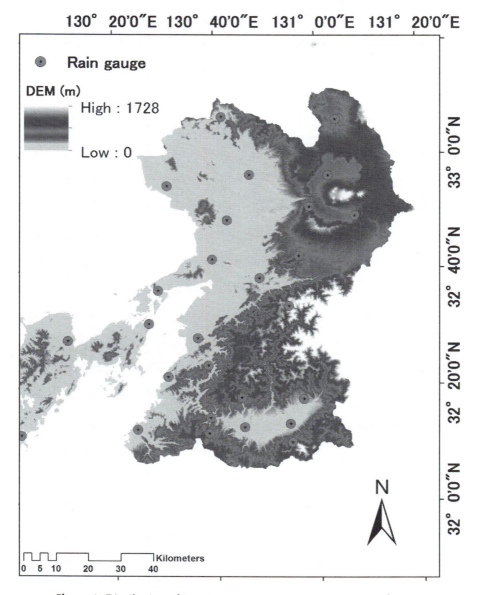

Figure 1 Distribution of 27 rain gauge stations in Kumamoto Prefecture.

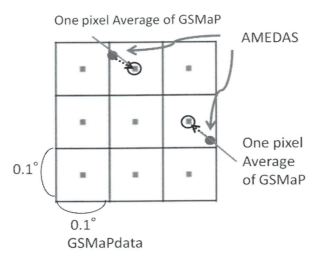

Figure 2 Point by point analysis.

TABLE 1 Detail product.

Product	Temporal resolution	Spatial resolution	Start date	Delay
GSMaP	1 hour	0.1 × 0.1 degree	March 2000	4 hour
Rain gauge	1 hour	Single point	November 1974	

Time Series Graph

A time series graph was conducted to estimate the rainfall pattern before the flood occurrence and to estimate the time lag between GSMaP precipitation data and observed rainfall data.

Verification Method

Visual Verification

There are several verification methods which can be used; one of them is visual verification. In this method, formatting the single point of rain gauge data sets into the spatial distributions with the same projection and the same color scale with GSMaP datasets was conducted. ArcGIS 10.1 was used as a tool to convert the single point data set into a raster data set, by using the kriging spatial interpolation method. The spherical model of kriging interpolation was chosen because of the very high correlation coefficient (0.9) with the observed rain gauge data. This method was used to convert the daily point gauge observed rainfall data to a 0.1 degree latitude/longitude grid. Gridded precipitation data from the ground station was used for visual comparison with GSMaP precipitation data.

Continuous Statistics

The aim of this method was to measure the correspondence between value of the estimates and the observation. To quantify the correspondence value, the following five statistical indices were used (Jiang et al. 2010). The correlation coefficient (r) was used to measure the fitness between GSMaP precipitation data and rain gauge observations which are shown in equation 1 (eq. 1). The Root Mean Square Error (RMSE) measured the average error magnitude (eq. 2). The mean absolute error (MAE) measured the average difference between GSMaP precipitation data and observed values (eq. 3). The relative bias (B) described the systematic bias of the satellite precipitation (eq. 4). The Nash-Sutcliffe (C_{NS}) measured the consistency of the satellite precipitation and gauge observation, both amount and temporal distribution (eq. 5). These indices are given by following equations.

$$r = \frac{\sum_{i=1}^{n}(G_i - \bar{G})(S_i - \bar{S})}{\sqrt{\sum_{i=1}^{n}(G_i - \bar{G})^2}\sqrt{\sum_{i=1}^{n}(S_i - \bar{S})^2}} \tag{1}$$

$$\text{RMSE} = \sqrt{\frac{1}{n}\sum_{i=1}^{n}(S_i - G_i)^2} \tag{2}$$

$$\text{MAE} = \frac{1}{n}\sum_{i=1}^{n}|S_i - G_i| \tag{3}$$

$$B = \frac{\sum_{i=1}^{n}(S_i - G_i)}{\sum_{i=1}^{n}G_i} \times 100\% \tag{4}$$

$$C_{NS} = 1 - \frac{\sum_{i=1}^{n}(S_i - G_i)^2}{\sum_{i=1}^{n}(G_i - \bar{G})^2} \tag{5}$$

where n is the total number of the rain gauge data or GSMaP data; S_i is the satellite estimates and G_i is the rain gauge observation values.

Categorical Statistics

Categorical statistics are used to measure the correspondence between the estimated and observed occurrence of events. Two categorical statistics were used, namely, the probability of detection (POD) and the false alarm ratio (FAR). POD measured how often the rain occurrence was correctly detected by satellite. FAR represented the fraction of diagnosed events that turned out to be wrong. Table 2 summarizes the contingency to assess GSMaP rainfall detection capability with rain or no rain events. The threshold of rain/no rain used in the contingency table is 0 mm/hour. In Table 1, "hits" represents correctly estimated rain events, "misses" describes when rain is not estimated but actual rain occurs, "false alarm" represents when rain is estimated but actual rain doesn't occur and "correct negative" represents correctly estimated no-rain events. Using the results shown in Table 2, the parameters POD and FAR are calculated by following equations.

$$POD = \frac{hits}{hits + misses} \tag{1}$$

$$FAR = \frac{false\ alarm}{hits + false\ alarm} \tag{2}$$

TABLE 2 Contingency table of yes or no events/with rain or no rain.

Observed rainfall	Estimated rainfall	
	Yes	No
Yes	hits	misses
No	false alarm	correct negative

○ RESULT AND DISCUSSION

History of Floods in Kumamoto Prefecture

Between 2003 and 2013, there were nine flood eventsoccurred in Kumamoto Prefecture. Flood mostly occurs in June or July, which is the rainy season in Japan. The floods spread mostly in the low altitude as shown in Fig. 3, except in Asootohime. During that period, Yamato city was hit by flood two times (2006 and 2007) and the highest flood frequency was in 2007 (four times). The floods occurred when the heavy rain was ranging from 221 mm/week to 608 mm/week. Extreme rainfall occurred in 2003 and 2012 and caused flash flooding and landslides. As a result, in 2012, 28 deaths were reported from this event with thousands forced to evacuate fast and widespread property damage.

Time Series Graph

There were two kinds of rainfall pattern which caused flood, namely, a long term pattern and a short term pattern. The long term pattern refers to accumulative rainfall from one day to several days causing flood. The short term rainfall pattern refers to accumulative rainfall for several hours causing flood.

Long Term Pattern

Figs. 4a and b shows the rainfall pattern before the flood occurrence in Gyuto city. The black line indicates hourly observed rainfall by AMEDAS while the orange line indicates hourly rainfall estimates by GSMaP_MVK. The pattern of both rainfall data was similar, but there was a time lag of 9 hours. From this result, it seems that GSMaP_MVK was able to predict the rainfall occurrence around Kumamoto Prefecture for flood monitoring because GSMaP_MVK can observe the amount of rainfall nine hours earlier.

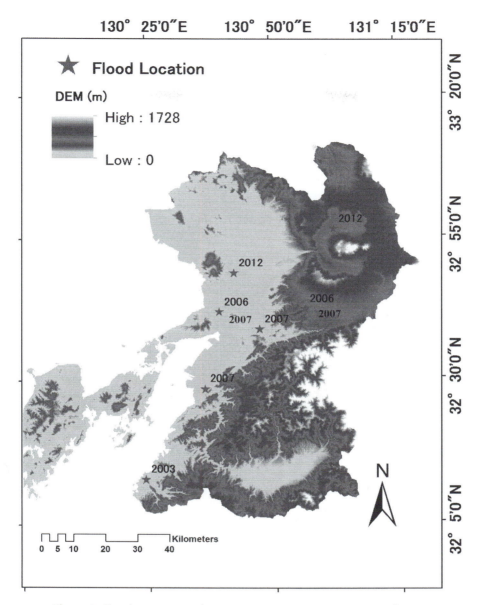

Figure 3 Flood occurrence during 2003 to 2013 in Kumamoto Prefecture.

Fig. 4b shows the rainfall pattern for AMEDAS data and GSMaP_MVK data after time lag calculation. In general, rainfall data from GSMaP_MVK were lower than rain gauge data especially during extreme rainfall (i.e. the highest rainfall amount of AMEDAS data was 50 mm/hr while in the same time the GSMaP_MVK data was 9 mm/hr). In addition, the total amount of rainfall which trigger the flash flood in the Gyuto city was 422 mm/week. This result shows that the rainfall pattern before the flood occurred was almost same and is classified as long term rainfall period because it took several days to trigger flash flood. Moreover, long term rainfall pattern occurred in 2007 and 2006 in Yatsushiro city, Yamato city, Mifune city, Misato city and Misato city. In those events, the total amount of rainfall in a week ranged from 406 mm to 608 mm.

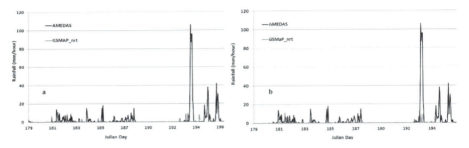

Figure 4 Rainfall pattern before the flood occurred in Gyuto city, 25 July 2006 (a); and after time lag matching (b).

Short Term Pattern

The most recent flood occurred in Kumamoto Prefecture was on 12 July 2012. GSMaP_MVK is only available from 2000 to 2010, therefore GSMaP_NRT was chosen as a satellite precipitation data in this research. Both GSMaP_NRT and GSMaP_MVK have same algorithm, but GSMaP_MVK is the reanalysis version of GSMaP_NRT (EORC & JAXA 2013a). Fig. 5a shows the rainfall pattern before the flood occurrence in Kumamoto city. The black line indicates hourly observed rainfall by AMEDAS while the orange line indicates hourly rainfall estimates by GSMaP_nrt. The pattern between AMEDAS data and GSMaP data is hard to recognize, but the rainfall occurrence can be detected. The GSMaP_NRT value underestimated rainfall and the time lag with AMEDAS data was 9 hours earlier. GSMaP_NRT still has a possibility to detect the rainfall occurrences as explained later. The time series graph after time lag matching is shown in Fig. 5b.

Fig. 5b shows the rainfall pattern of two weeks difference and looks similar in the 12 July 2012. A short term rainfall pattern occurred in this city. On 12 July 2012, heavy rainfall occurred for 5 hours with the peak rainfall amount of 435 mm for 5 hours. Asootohime at high altitude had heavy rainfall of 435.5 mm for 5 hours. As a consequence, a flash flood hit Asootohime city for the first time after 30 August 1980. This disaster caused 25 people dead/missing and 385 destruction of house and public facility (Yokota et al. 2014). In addition, the short term rainfall pattern had a rainfall amount ranging from 199 mm to 435 mm for 5 hours. Because of this pattern, measurement of high temporal resolution of precipitation data became very important. However, the uncertainty precipitation is the main source of uncertainty impacting on the flood forecasts (Nester et al. 2012). Thus, the several approaches should be used to measure rainfall characteristics. One approach is to use precipitation satellite data which is easy to get, has high temporal resolution data and can easily reach isolated areas. For that reason, verification of hourly precipitation satellite data is necesssary.

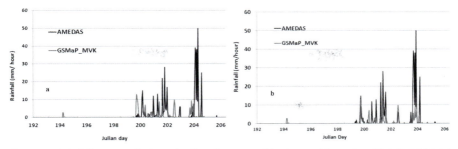

Figure 5 Rainfall pattern before the flood occurred in Asootohime city, 12 July 2012 (a); and after time lag matching (b).

⟶ VISUAL VERIFICATION ═══════════════

Fig. 6 shows visual verification of GSMaP_MVK data after 9 hours time lag matching. It shows that the spatial distribution was almost same in the northern part of Kumamoto Prefecture when the light rain came. In contrast, the spatial distribution difference occurred when heavy rain comes (i.e heavy rains was recorded in the eastern part of the study area by AMEDAS, but it was recorded in the southeastern part of the sudy area by GSMaP_MVK). However, the GSMaP_MVK data still underestimated the actual rainfall. Fig. 7 shows visual verification of GSMaP_NRT data after 9 hours time lag matching. It shows that the spatial distribution was different and that the GSMaP_NRT data also underestimated actual rainfall. The highest concentration of GSMaP_NRT data in Fig. 7 was between 10 to 15 mm/hr.

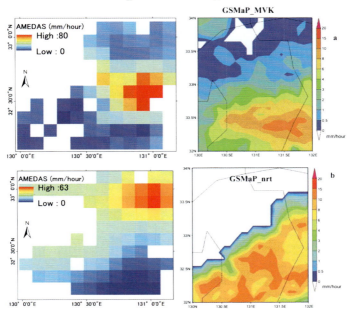

Figure 6 (a) Spatial distribution of observed rainfall 6 July 2007 at 09.00 (left) and rainfall estimates by GSMaP_MVK 6 July 2007 at 00.00 (right) (b) Spatial distribution of observed rainfall 12 July 2012 at 06.00 (left) and rainfall estimates by GSMaP_NRT 11 July 2012 at 22.00 (right).

Statistical Verification

GSMaP_MVK

Fig. 8 shows the validation result of GSMaP_MVK in Kumamoto Prefecture. The value of the five continuous statistics showed that GSMaP_MVK was reasonably good at detecting precipitation events before the flood occurrence. For the area averaged hourly rainfall, the correlation coefficient was 0.52, RMSE was 5.148 mm/hr with the bias percentage of –54.09% indicating an underestimation of rainfall. The underestimation of rainfall is consistent with the previous finding (Fukami et al. 2010; Kubota et al. 2009; Seto et al. 2008; Shrestha et al. 2011). Underestimation of GSMaP data resulted from no microwave radiometer information during the peak period for heavy rainfall (Kubota et al. 2009) and sudden increases in rain rate did not reflect the IR brightness temperature (Ushio et al. 2009). In Kumamoto Prefecture, hourly GSMaP_MVK data has a weak correlation coefficient compared with the previous study which validated daily GSMaP_MVK data (Makino 2012). Table 3 shows the result of categorical statistics for which 2312 total points were observed. Hits frequency was 446 times, misses frequency was 135 times, false alarm frequency was 277 times and correct negative frequency was 1454 times.

TABLE 3   Contingency table of yes or no events/with rain or no rain of GSMaP_MVK.

AMEDAS	GSMaP_MVK	
	Rain	**No rain**
Rain	446	135
No rain	277	1454

POD: 0.768; FAR: 0.383

Based on Table 3, POD and FAR value were 0.77 and 0.38. It means that 77% of rain occurrences were correctly detected and 38% of rain occurrences turned out to be wrong by GSMaP_MVK. These values of two categorical statistics showed that GSMaP_MVK product was reasonably good at detecting the precipitation events over Kumamoto Prefecture.

GSMaP_NRT

Fig. 8 shows the validation result of GSMaP_NRT in Kumamoto Prefecture. The value of the five continuous statistics showed that GSMaP_NRT was not good at detecting precipitation events before the flood occurred. For the area averaged hourly rainfall, the correlation coefficient was 0.24 RMSE was 8.272 mm with the bias percentage of –87.042% indicating underestimation of rainfall. Table 4 shows the result of categorical statistics which 716 total points were observed. Hits frequency was 76 times, misses frequency was 137 times, false alarm frequency was 37 times and correct negative frequency was 466 times.

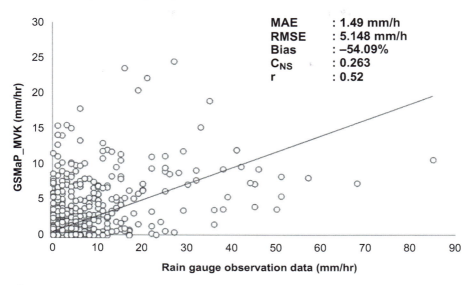

Figure 7 Scatter plot of rain gauge observation and GSMaP_MVK hourly scales in Kumamoto Prefecture.

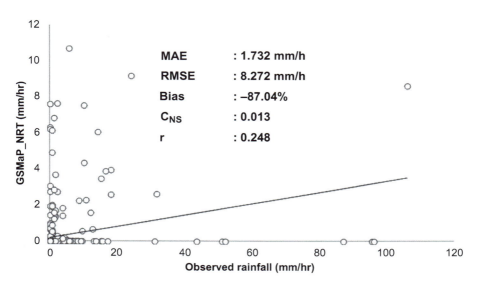

Figure 8 Scatter plot of rain gauge observation and GSMaP_NRT hourly scales in Kumamoto Prefecture.

TABLE 4 Contingency table of yes or no events/with rain or no rain of GSMaP_NRT.

AMEDAS	GSMaP_NRT	
	Rain	No rain
Rain	76	137
No rain	37	466

POD: 0.357; FAR: 0.327

Based on Table 4, POD and FAR values was 0.357 and 0.327. It means 35.7% of rain occurrences were correctly detected and 32.7% of rain occurrences turned out to be wrong by GSMaP_NRT. These values of two categorical statistics showed that GSMaP_NRT product was not so good at detecting the precipitation events in Kumamoto Prefecture. Nevertheless, GSMaP_NRT has the value of FAR similar with GSMaP_MVK and can be downloaded after 4 hour satellite observation. As a result, GSMaP_NRT can be used as emergency data analysis for precipitation data when rainfall observation data is not available.

○ CONCLUSIONS

There were two rainfall patterns over Kumamoto Prefecture before the flood occurrence, namely "the long term period" and "the short term period". In the long term period, the flood occurred when the rainfall amount ranged from 406 mm to 608 mm for one week, while the short term period was from 199 to 435 mm for five hours. GSMaP_MVK was reasonably good at detecting precipitation events before the flood occurrence both spatially and temporally. GSMaP_NRT was not good at detecting precipitation events before the flood occurrence, especially for spatial distribution, but it can be used as emergency data analysis for precipitation data when rainfall observation data is not available. For the future research, evaluation and bias correction of GSMaP data during rainy season is necessary to reduce the error, especially during heavy rainfalls.

○ ACKNOWLEDGEMENT

We would like to thank LPDP (Indonesia Endowmnet Fund for Education of Ministry of Finance) for financial support. Our deep appreciation to Prof. Anthony Hoysted for reviewing our paper. We also thank JAXA (Japan Aerospace Exploration Agency) and JMA (Japan Meteorological Agency) for GSMaP data and ground rainfall data.

○ REFERENCES

Aonashi, K., Shibata, A. and Liu, G. 1996. An over ocean precipitation retrieval using SSM/I multichannel brightness temperature. Journal of the Meteorological Society of Japan 74: 617-637.

As-syakur, A.R., Tanaka, T., Prasetia, R., Swardika, I.K. and Kasa, I.W. 2011. Comparison of TRMM multisatellite precipitation analysis (TMPA) products and daily-monthly gauge data over Bali. International Journal of Remote Sensing 32(24): 8969-8982.

EORC and JAXA. 2013a. User`s Guide for Global satellite mapping of precipitation microwave – IR Combined Product (GSMaP_MVK) version 2.5, Tsukuba, Japan, EORC & JAXA.

EORC and JAXA. 2013b. Global Rainfall Map in Near-Real-Time by JAXA Global Rainfall Watch (GSMaP_NRT) version 2.5. Tsukuba, Japan.

Fukami, K., Shirashi, Y., Inomata, H. and Ozawa. G. 2010. Development of integrated flood analysis system (IFAS) using satellite-based rainfall products with a self-correction method. International centre for water hazard and risk management under auspices of UNESCO (ICHARM). Public Works Research Institute. Tsukuba, Japan.

Jiang, S., Ren, L., Yong, B., Yang X. and Shi. L. 2010. Evaluation of high-resolution satellite precipitation products with surface rain gauge observations from Laohahe Basin in northern China. Water Science and Engineering 3: 405-417.

Kazama, S., Sato, A. and Kawagoe, S. 2009. Evaluating the cost of flood damage based on changes in extreme rainfall in Japan. Sustainability Science 4: 61-69.

Kobold, M. and Suselj, M. 2005. Precipitation forecast and their uncertainty as input into hydrological models. Hydrology and Earth System Sciences 9(4): 322-332.

Kubota, T., Ushio, T., Shige, S., Kida, S., Kachi, M. and Okamoto, K. 2009. Verification of high resolution satellite-based rainfall estimates around Japan using a gauge calibrated ground radar data set. Journal of the Meteorological Society of Japan 87A: 203-222.

Makino, S. 2012. Verification of the accuracy of rainfall data by global satellite mapping of precipitation (GSMaP) Product, Master Thesis, Yamaguchi University, Yamaguchi, Japan.

Nester, T., Komma, J., Vigliore, A., Bloschl, G. 2012. Flood forecast error and ensemble spread-case study. Water Resources Research 48: w1052.

Okamoto, K., Iguchi, T., Takahashi, N., Ushio, T., Awaka, J., and Kozu, T. et al. 2007. High precision and high resolution global precipitation map from satellite data. ISAP 2007, Niigata, Japan.

Pauwels, V.R.N. and Lannoy, G.J.M. 2005. Improvement of modeled soil wetness conditions and turbulent fluxes through the assimilation of observed discharge. Journal of Hydrometeorology 7: 458-476.

Prasetia, R., As-syakur, A.R., Osawa, T. 2013. Validation of trmm precipitation radar satellite data over Indonesia region. Theory and Application of Climatology 112: 575-587.

Seto, S. 2009. An evaluation of overland rain rate estimates by the GSMaP and GPROF Algorithm: the role of lower frequency channels. Journal of the Meteorological Society of Japan (87A): 183-202.

Shrestha, M.S., Takara, K., Kubota, T. and Bajracharya. S.R. 2011. Verification of GSMaP rainfall estimates over the central Himalayas. Journal of Hydrologic Engineering 67: I37-I42.

Ushio, T., Sasashige, K., Kubota, T., Shige, S., Okamoto, K., Aonashi, K. et al. 2009. A Kalman filter approach to the global satellite mapping of precipitation (GSMaP) from combined passive microwave and Infrared radiometric data. Journal of the Meteorological Society of Japan (87A): 137-151.

Yokota, I., Nishiyama, K., Tsukahara, K. and Wakimizu, K. 2014. Visually and easily detectable radar information available for judging the evacuation from heavy rainfall disasters. Proc. The 8th ERAD Germany. 1-6.

10

CHAPTER

◇◇

Appraisal of Surface and Groundwater of the Subarnarekha River Basin, Jharkhand, India: Using Remote Sensing, Irrigation Indices and Statistical Technique

Sandeep Kumar Gautam,[1] *Abhay K. Singh,*[2] *Jayant K. Tripathi,*[1] *Sudhir Kumar Singh,*[3,]* *Prashant K. Srivastava,*[4] *Boini Narsimlu*[5] and *Prafull Singh*[6]

ABSTRACT

The qualitative assessment of surface and groundwater was carried out of the Subarnarekha River Basin (SRB), Jharkhand in the year 2008. The specific objective of the study was to find out the irrigational and drinking suitability of surface and groundwater. The land use/land cover (LULC) map of the study area was derived from Landsat TM of the year 2008. LULC statistics shows that cropland has an area 12555.99 km^2 (47.31%), forest area is 13440.55 km^2 (50.64%), water body area is around 316.15

[1] School of Environmental Sciences, Jawaharlal Nehru University, New Delhi, India.
[2] Central Institute of Mining and Fuel Research, Barwa Road, Dhanbad, Jharkhand, India.
[3] K. Banerjee Centre of Atmospheric and Ocean Studies, IIDS, Nehru Science Centre, University of Allahabad, Allahabad (U.P.), India.
[4] Institute of Environment and Sustainable Development, Banaras Hindu University, Varanasi, India.
[5] Division of FM and PHT, ICAR–Indian Grassland and Fodder Research Institute, Jhansi (U.P.), India.
[6] Amity Institute of Geo-Informatics and Remote Sensing, Amity University, Noida, India.
* Corresponding author: sudhirinjnu@gmail.com

km^2 (1.19%) and built up area is around 224.53 km^2 (0.84%). The average concentration of pH, EC (μS cm^{-1}), TDS, F$^-$, Cl$^-$, HCO$_3^-$, SO$_4^{2-}$, NO$_3^-$, dissolved silica, Ca^{2+}, Mg^{2+}, Na$^+$ and K$^+$ are 7.61, 207.33 (μS cm^{-1}), 170.95, 0.51, 13.53, 87.98, 11.62, 3.69, 15.17, 19.29, 6.33, 10.05 and 2.84 mgL^{-1} in surface water samples, respectively, while for the same, the average concentration it is 7.21, 965.19 (μS cm^{-1}), 805.53, 0.56, 145.49, 254.79, 109.49, 46.80, 38.58, 133.16, 34.26, 39.21 and 3.21 mgL^{-1} in groundwater samples, respectively. Sodium absorption ratio (SAR), sodium percentage (Na%), residual sodium carbonate (RSC), permeability index (PI), magnesium hazard (MH) and Kelly's Index (KI) values of surface and groundwater show that most of water samples are suitable for irrigation purposes. However, the water samples categorized on the Piper diagram suggest that all the surface water samples are of Ca–HCO$_3$ type. Moreover, 47.37% groundwater samples fall in Ca-Cl type, 36.84% fall in Ca-HCO$_3$ type, 10.53% fall in Ca-SO$_4$ type and remaining 5.26% groundwater sample fall in Mg-HCO$_3$ type. The clustering was performed to find similar sites in both the seasons.

KEYWORDS: Surface water; groundwater; hydrogeochemistry; water quality; land use; satellite data.

○ INTRODUCTION

Water pollution in developing countries increases after industrialization, unprecedented population growth, urbanization after the globalization era, i.e. 1990 onwards (Singh et al. 2013a). Consequently, surface and groundwater are also contaminated and affected by many factors such as the expansion of irrigation activities, industrialization and urbanization (Krishna et al. 2009; Foster et al. 2002). Thus man has manipulated the hydrological cycle both quantitatively and qualitatively by interfering with the natural processes at various steps either through addition of solid or liquid materials or through withdrawal of water, sediments etc. Quality of water in India is deteriorating at very fast rate (Singh et al. 2012; Singh et al. 2015). Natural and anthropogenic sources along with chemical and biogeochemical constituents have been considerably altering surface and groundwater quality in recent years (Chidambaram et al. 2013). It has become very important to understand the chemical composition of water to understand the anthropogenic and geogenic impact on the water sources for better management and to develop appropriate treatment technologies (Gautam et al. 2013). Hence, monitoring and conservation of water resources are highly requisite for a sustainable environment and to fulfill the fresh water demand (Sun et al. 1992).

Remote sensing and Geographic Information System (GIS) are very useful tools which could be used for synoptic representation of water quality of any area (Gupta and Srivastava 2010). Land use/land cover (LULC) changes quantification through satellite remote sensing is one of the major applications; it is important for assessing global environmental change processes, supports in making policies and optimizes the use of natural resources (Srivastava et al. 2012a). The LULC types, such as agricultural land and urban area associated with human activities, often affect both the surface and groundwater quality. Hence monitoring spatial-temporal changes is essential to understand the driving factors that influence the water quality of an area (Merchant 1994; Wu and Segerson 1995; Srivastava et al. 2013). On the other hand, GIS is an important tool for spatial analysis and integration of spatial and non-spatial data to derive useful outputs (Singh et al. 2013a) and helps in decision making. It can be used for formulating a simple and robust water quality pollution assessment tool for rapid information generation and broadcasting to water resources managers and the public (Vasanthavigar et al. 2010; Singh et al. 2012).

The application of multivariate statistical method is very useful for classification, modeling and interpretation of large data sets which allow the reduction of dimensionality of the large data sets (Singh et al. 2005; Singh et al. 2009). The present study applied the above-mentioned methodologies, by using different valid standard techniques such as primary data (water quality) generation for the geologically heterogeneous area following the standard laboratory procedure and multivariate analysis in correspondence with remote sensing and GIS techniques. Therefore, the foremost objective of this research focuses on the utilization of different irrigation indices to evaluate the status of water of SRB.

○ MATERIALS AND METHODS ════════════════════════

Study Area

Subarnarekha River is one of the major rivers of the south Chotanagpur plateau of Jharkhand state in India. It is a rain fed river and originates from the Nagri village of Ranchi plateau (23°.4′N, 85°.4′E) situated at 756 m above mean sea level (Fig. 1). With a total length of 470 km the river passes through states like Jharkhand, West Bengal and Orissa and finally falls into the Bay of Bengal. Its total catchment area comprises 18,950 km², of which 13,590 km² are situated in Jharkhand, 3200 km² in Orissa and 2160 km² in West Bengal (Rao 1995). The dominant LULC includes agricultural and forest land. The basin is famous for mining activities. Mineral resources are associated with a Precambrian shear zone (Négrel et al. 2007) and contained copper and uranium. It is an asymmetrical catchment basin, on the right bank there are four major tributaries, the Raru, the Kanchi, the Karkai and the Kharkai and on the left side only one i.e. the Dulung. An important tributary of this river is the Kharkai that originates from the neighboring Mayurbhanj district and merges with Subarnarekha River near the city Jamshedpur.

Physiography and Climate

The SRB occupies a region of varied physiography ranging from steep hill masses to flat coastal plains through a series of dissected plateaus and sloping plains. The following six physiographic divisions in the SRB have been recognized as Ranchi plateau, escarpments and plateau slopes, uplands, central plains, intervening hill ranges and coastal plains (Mukhopadhyay 1980). The climate is dry sub-humid to humid sub-arid types. The cold winter season (November to February) has temperature between 10°C to 20°C. The hot summer (March to mid-June) has temperature between 37°C to 20°C. The average annual rainfall is 1400 mm, around

82.1% rainfall is received from the months of June to September and the remaining 17.9% during the rest of the months.

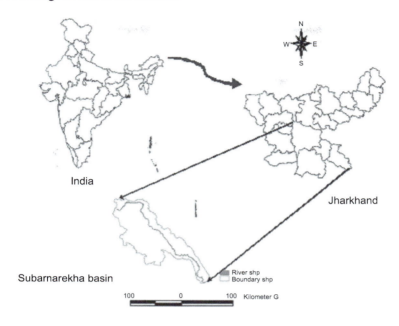

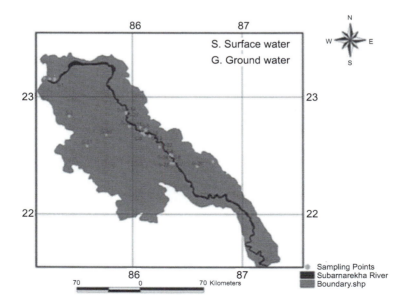

Figure 1 Location map and sampling stations in Subarnarekha River Basin (SRB).

Geology of SRB

The SRB lies on the Indian shield where ancient Precambrian igneous and metamorphic rocks are exposed. It is only in the lower reaches of the basin, southeast of Ghatsila, that the younger geological formations, mainly tertiary gravels, Pleistocene alluvium and recent alluvium are exposed. SRB is situated under by folded and fractured Precambrian metasediments, mainly mica schists, quartzite and hornblende schists (Ghosh et al. 2002). Only fracture type aquifers are identified and groundwater occurs under unconfined condition in the hard rock areas (Négrel et al. 2007). It is a part of the Precambrian terrain of the Singhbhum craton in the eastern part of India. The prominent 200 km long copper belt thrust zone of Singhbhum separates the Precambrian basement of the region into two distinct provinces i.e. the Singhbhum-Orissa iron ore province on the south and the Satpura province on the north. The craton consists of eight principal lithological associations (Ramakrishnan and Vaidyanadhan 2008): (i) Singhbhum granite with enclaves of Older Metamorphic Tonalite Gneiss (OMTG) and Older Metamorphic Group (OMG) of sediments and volcanics, (ii) basins of Banded iron formation (BIF) around the Singhbhum granite, (iii) Volcanic basins, loosely termed as greenstone belts, (iv) Flysch-like sediments and volcanics of North Singhbhum orogeny, (v) Mafic dyke swarms, (vi) Kolhan basin, (vii) newer tertiary and (viii) alluvium. The SRB is rich in minerals resources like ores of copper, iron, uranium, chromium, gold, vanadium, limestone and dolomite, kyanite, asbestos, barytes, apatite, china clay, talc and steatite, building stones etc. Fig. 2 shows the geological map of the study area (Gautam et al. 2015).

Remote Sensing and GIS Implementation

LULC classification of the area has been estimated on ENVI v4.8 platform using the Landsat TM satellite data of year 2008. Maximum Likelihood Classification (MLC) a supervised classification approach is employed to determine the LULC. Arc GIS (9.3) software is used for the preparation of primary thematic layers. The image projection parameters are used as Universal Transverse Mercator, World Geodetic System 84 (WGS84). MLC is expressed as by equation 1.

$$D = \ln(a_c) - [0.5 \ln(|cov_c|)] - [0.5 (X - M_c) T(cov_c - 1) (X - M_c)] \tag{1}$$

where, D is weighted distance; c is a particular class; X is the measurement vector of the particular pixel; M_c is the mean vector of the sample of class; a_c is percent probability that any particular pixel is a member of class c; (Defaults to 1.0); Cov_c is the covariance matrix of the pixels in the sample of class c; $|Cov_c|$ is determinant of Cov_c; $Cov_c - 1$ is inverse of Cov_c; ln is natural logarithm function; $T =$ transposition function.

For the validation of classes in the classified image, the accuracy assessment is performed. In order to evaluate the performance of the MLC classifiers, the accuracy assessment of the classified image is provided through the ground control points (GCPs), assuring distribution in a rational pattern so that a specific number of observations are assigned to each category on the classified image. The Kappa accuracy was computed, as expressed in equation 2 (Bishop et al. 1975).

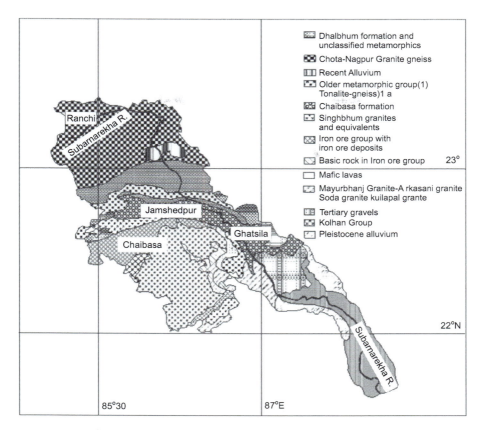

Figure 2   Geology map of Subarnarekha River basin (SRB).

$$\kappa = \frac{N \sum\limits_{i=1}^{r} X_{ii} - \sum\limits_{i=1}^{r} (x_{i+})(x_{+i})}{N^2 - \sum\limits_{i=1}^{r} (x_{i+})(x_{+i})} \tag{2}$$

where, r is the number of rows in the matrix, X_{ii} is the number of observations in row i and column i (the diagonal elements), x_{+i} and x_{i+} are the marginal totals of row r and column i, respectively and N is the number of observations.

Sample Collection and Physico-chemical Analysis

Table 1 illustrates the sampling locations and station codes. Suspended sediments were separated from the water samples in the laboratory by using 0.45 μm Millipore membrane filters of 47 mm diameter. pH of collected water sample was measured by Consort microcomputer ion meter. Electrical conductivity was measured by using Consort pH and Conductivity Meter. It provides measurement of electrical conductivity by a cell consisting of two platinum electrodes to which an alternating potential is applied. Bicarbonate was determined by potentiometric titration method. The dissolved silica was determined by molybdosilicate method (APHA 2005). Major anions (F^-, Cl^-, NO_3^- and SO_4^{2-}) were analysed through ion chromatograph (Dionex Dx-120) using anions AS12A/AG12 columns coupled to an anion self-regenerating suppressor (ASRS) in recycle mode. Major cations (Ca^{2+}, Mg^{2+}, Na^+ and K^+) were measured by ion chromatograph by using cation column (CS12A/CS12G) and cation self-regenerating suppressor (CSRS) in recycle mode. Analysis of major cations of all samples was also repeated on ICP-AES (Jobin-Yvon, ULTIMA 2) to monitor the accuracy of the analysis. The analytical precision was maintained by running the known standard after every 10 samples. An overall precision, expressed as percent relative standard deviation (RSD) was obtained for the entire samples. Analytical precision for cations (Ca^{2+}, Mg^{2+}, Na^+ and K^+) and anions (F^-, Cl^-, NO_3^- and SO_4^{2-}) were within 10%.

TABLE 1 The sampling location, station code and water type.

S. No.	Locations	Sample code	Water type	Latitude	Longitude
1	Hatia bridge	1S	Ca-HCO$_3$	N23°16.977′	E085°18.557′
2	Sonari	2S	Ca-HCO$_3$	N22°50.156′	E086°09.765′
3	Chandil	3S	Ca-HCO$_3$	N22°58.278′	E086°01.217′
4	Maubhandar	4S	Ca-HCO$_3$	N22°35.562′	E086°26.782′
5	Mango	5S	Ca-HCO$_3$	N22°48.998′	E086°12.661′
6	Ghatsila	6S	Ca-HCO$_3$	N22°34.896′	E086°28.215′
7	Musabani	7S	Ca-HCO$_3$	N22°30.468′	E086°29.117′
8	Kanchi, Khuddi	8S	Ca-HCO$_3$	N23°10.754′	E085°16.628′
9	Kharkai, Adityapur	9S	Ca-HCO$_3$	N22°47.316′	E086°10.437′
10	Kharkai, Sonari	10S	Ca-HCO$_3$	N22°50.131′	E086°09.610′
11	Sankh, Pampughat	11S	Ca-HCO$_3$	N22°30.442′	E086°29.090′
12	Durwa Dam	12S	Ca-HCO$_3$	N23°17.648′	E085°15.532′
13	Chandil Dam	13S	Ca-HCO$_3$	N22°58.564′	E086°01.430′
14	Hatia bridge	14G	Ca-Cl	N23°17.210′	E085°18.584′
15	Khuddi village	15G	Ca-HCO$_3$	N23°10.731′	E085°16.596′
16	Tatanagar railway station	16G	Ca-HCO$_3$	N22°46.344′	E086°11.731′

Table 1 Contd.

S. No.	Locations	Sample code	Water type	Latitude	Longitude
17	Adityapur	17G	Ca-Cl	N22°47.306′	E086°10.322′
18	Sakchi	18G	Ca-Cl	N22°48.421′	E086°12.373′
19	Govindpur	19G	Ca-SO$_4$	N22°45.537′	E086°15.101′
20	Mango	20G	Ca-Cl	N22°49.113′	E086°12.706′
21	Jugsalai	21G	Ca-Cl	N22°46.525′	E086°11.281′
22	Jaduguda	22G	Ca-HCO$_3$	N22°39.663′	E086°20.829′
23	Mushabani	23G	Ca-Cl	N22°30.713′	E086°27.443′
24	Maubhandar	24G	Ca-SO$_4$	N22°35.47′	E086°26.703′
25	Ghatsila	25G	Mg-HCO$_3$	N22°34.896′	E086°28.215′
26	Chakulia	26G	Ca-HCO$_3$	N22°28.837′	E086°43.253′
27	Chandil	27G	Ca-HCO$_3$	N22°58.370′	E086°01.429′
28	Kandra	28G	Ca-Cl	N22°51.009′	E086°03.024′
29	Saraikela	29G	Ca-Cl	N22°42.125′	E085°55.905′
30	Chaibasa	30G	Ca-Cl	N22°46.420′	E085°48.968′
31	Charadharpur	31G	Ca-HCO$_3$	N22°40.376′	E085°37.885′
32	Chakulia	32G	Ca-HCO$_3$	N22°28.749′	E086°43.252′

S: surface water; G: groundwater

Statistical Analysis

Cluster analysis (CA) technique is applied; it groups the objects (cases) into classes (clusters) on the basis of similarities/dissimilarities within or between classes respectively. CA is used to see the pattern in the data sets implemented using hierarchical agglomerative clustering technique by means of the Euclidean distances following Ward's method (Singh et al. 2009; Srivastava et al. 2012b).

○ RESULTS AND DISCUSSIONS ══════════════

Land Use/Land Cover

The overall accuracy of the classified image is >91%. The classified image (Fig. 3) shows overall that the study area is predominantly agricultural land and forest land and mixed forest land, with less than 2 percent of the total land area dedicated to water bodies (Table 2). The LULC conversion is taking place in SRB which may be probably polluting its water resources. Cropland has an area 12555.99 km^2 (47.31%), forest is 13440.55 km^2 (50.64%), water body area is around 316.15 km^2 (1.19%) and built up area is around 224.53 km^2 (0.84%).

TABLE 2 Land use/land cover statistic in the year 2008.

S. No.	Land use/land cover classes	Area (km²)	Area in percent
1	Water body	316.15	1.19
2	Forest	13440.55	50.64
3	Built up	224.53	0.84
4	Cropland	12555.99	47.31
5	Total	26537.23	100

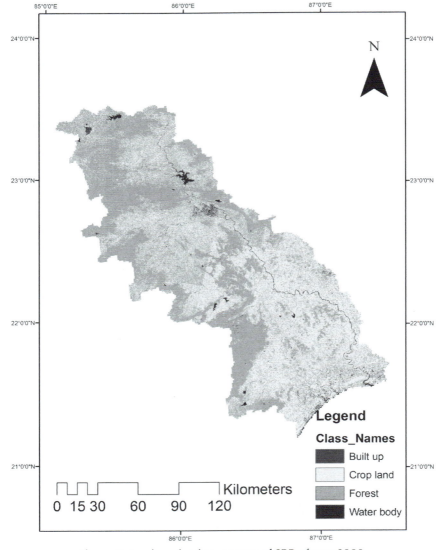

Figure 3 Land use/land cover map of SRB of year 2008.

The groundwater sampling sites at the elevated locations will generally have the safer water supply. The farthest downslope would receive the combined effluent from the other sources, industries as well as from agricultural fields and forest because the liquid effluent follows the same path as the surface runoff and hence contaminated water is found more toward downward slopes (Waller 2001). In general, the sites near agricultural fields and forests have more nitrate concentration. The sampling sites which fall in the urban area have higher a concentration of nitrate and sulphate due to sewage.

Hydrogeochemistry of Surface/Groundwater

The descriptive statistics of 13 physico-chemical parameters at 32 locations are given in Table 3. The mean pH values of the 13 surface water sampling locations showed very little variation ranging from 7.07 to 7.87 (average 7.61) while the groundwater sampling locations at 19 stations showed higher variation, ranging from 5.97 to 7.67 (average 7.21) respectively, implying slight alkaline in surface and groundwater. The highest value of pH is 7.87 and 7.67 reported in Maubhandar (4S) and Adityapur (17G) in surface and groundwater samples, respectively. The lowest value of pH is 7.07 and 5.97 reported in Hatia Bridge (1S) and Chakulia (26G) in case of surface and groundwater samples. This may be the result of the industrial and agricultural activities in the region. The mean electrical conductivity is 207 and 965 μS cm^{-1} in surface and groundwater samples. Highest EC 290 μS cm^{-1} is reported at Ghatsila (6S) in surface water samples and 2833 μS cm^{-1} at Govindpur (19G) in groundwater samples. The lowest EC 74 μS cm^{-1} is reported at Durwa Dam (12S) in surface water samples and 304 μS cm^{-1} at Chakulia (32G). The mean TDS values are 171 and 806 mgL^{-1} in surface and groundwater sample locations, respectively. The highest values of TDS are 238 and 2319 mgL^{-1} at Ghatsila (6S) and Govindpur (19G) in surface and groundwater samples respectively. The lowest values of TDS are 57 and 226 mgL^{-1} at Durwa Dam (12S) and Chakulia (32G) in surface and groundwater samples. The mean fluoride value is 0.51 and 0.56 mgL^{-1} in surface and groundwater sample locations respectively. The highest value of fluoride in the surface and groundwater sample is 0.92 mgL^{-1} at Musabani (7S) and 1.50 mgL^{-1} at Ghatsila (25G). The lowest value of fluoride in the surface and groundwater sample is 0.24 mgL^{-1} at Sankh, Pampughat (11S) and 0.12 mgL^{-1} at Hatia Bridge (14G) locations. The common geogenic sources of fluoride are fluorite (CaF_2) and apatite [$Ca_5(Cl, F, OH)(PO_4)_3$] minerals which occur in igneous and sedimentary rocks (Berner and Berner 1987). The mean chloride values are 13.53 and 145.49 mgL^{-1} in surface and groundwater sample locations and the highest value of chloride in surface water is 29.80 mgL^{-1} at Maubhandar (4S) and 390.37 mgL^{-1} at Saraikela (29G) in groundwater. The key natural sources of chloride in water are sea salt and dissolution of halite (NaCl) during weathering and anthropogenic sources are domestic and industrial sewage, mining and road dust. The mean values of bicarbonate are analyzed as 87.98 and 254.79 mgL^{-1} in surface and groundwater samples. The highest value of bicarbonate is 111.50 mgL^{-1} at Ghatsila (6S) in surface water and 443.03 mgL^{-1} at Govindpur (19G) in the groundwater sample. Bicarbonate in the water is mainly contributed to the water system from weathering reactions and decomposition of organic matter (Jha et al. 2009). The mean values of sulfate are 11.62 and 109.49 mgL^{-1} in the surface and groundwater sample and the highest value is 28.13 mgL^{-1} at Ghatsila (6S) in the surface water and 631.37 mgL^{-1} at Govindpur (19G) in groundwater sample, respectively. The mean values of nitrate are 3.69 and 46.80 mgL^{-1} in surface and groundwater samples with highest values of 7.63 mgL^{-1} at Ghatsila (6S) in surface water and 223.97 mgL^{-1} at Govindpur (19G) in groundwater sample.

Sulfur is widely distributed in reduced form in igneous and sedimentary rocks. The two major forms of sulfur in sedimentary rocks are sulfide sulfur in pyrite (FeS) and sulfate sulfur as gypsum ($CaSO_4.H_2O$) and anhydrite ($CaSO_4$) (Berner and Berner 1987). Dissolved silica in water system comes entirely from silicate weathering. The mean values of dissolved silica are 15.17 and 38.58 mgL^{-1} in surface and groundwater samples with highest values 22.73 mgL^{-1} at Sankh, Pampughat (11S) in surface water and 53 mgL^{-1} at Govindpur (19G) in groundwater sample. The mean values of calcium are 19.29 and 133.16 mgL^{-1} in surface and groundwater samples and highest values of 27.17 mgL^{-1} was recorded at Kharkai, Sonari (10S) in surface water and 376.07 mgL^{-1} at Govindpur (19G) in groundwater sample. The mean values of magnesium are 6.33 and 34.25 mgL^{-1} in surface and groundwater samples respectively and highest value is 9.30 mgL^{-1} recorded at Sankh, Pampughat (11S) in the surface water sample and 139.97 mgL^{-1} at Govindpur (19G) in the groundwater sample. Calcium and magnesium are contributed to surface and groundwater from rock weathering. The sources of Ca^{2+} mainly from carbonate rocks containing calcite ($CaCO_3$) and dolomite [$CaMg(CO_3)_2$] with a smaller amount derived from Ca-silicate minerals, chiefly calcium plagioclase and a minor fraction from $CaSO_4$ minerals. The main source of magnesium in surface and groundwater are Mg-silicate minerals, mainly amphiboles, pyroxenes, olivine and biotite, as well as dolomite (Berner and Berner 1987). The mean values of sodium are 10.05 and 39.21 mgL^{-1} in surface and groundwater samples with the highest value of 17.37 mgL^{-1} at Ghatsila (6S) in surface water and 105.33 mgL^{-1} at Govindpur (19G) in the groundwater sample. Sodium in silicate rock is present mainly as the albite component of plagioclase ($NaAlSi_3O_8$). Plagioclase is a major source of Na^+ for groundwater and thus it is also a major source for surface water. Potassium in water comes predominantly ($\sim 90\%$) from the weathering of silicate minerals, particularly K-feldspar (orthoclase and microcline) and mica (Berner and Berner 1987). The mean values of potassium are 2.84 and 3.21 mgL^{-1} in surface and groundwater samples with highest values of 5.03 mgL^{-1} at Musabani (7S) and 12.53 mgL^{-1} at Hatia Bridge (14G).

The elevated concentration of NO_3^- in groundwater was found at Govindpur, Hatia Bridge, Kandra, Musabani, Saraikela, Mango and Tatanagar railway station. The sources of NO_3^- in water are industries, mining, domestic wastes (septic systems, sanitary sewage effluent releases, domestic animal wastes) and excessive use of pesticides, manure and fertilizers applied in agriculture fields (Gautam et al. 2013). In the Govindpur location, concentration of TDS, Ca^{2+}, Mg^{2+}, SO_4^{2-} and NO_3^- were found above their prescribed limits defined by the Bureau of Indian Standards (BIS 2004-05) for drinking purposes. Therefore, proper management and remediation strategies should be taken before consumption of groundwater in Govindpur area. In

Jamshedpur (Tatanagar) district, effluents of a large number of industrial units, e.g., fertilizer (ammonium sulphate), inorganic chemicals (sulphuric acid), iron and steel and general engineering have polluted the surface and groundwater. This reflects anthropogenic and industrial influence on the geochemistry of local groundwater. The SRB is rich in mineral based industries mainly uranium in Jaduguda and copper mines present in Mushabani and Maubhandar (Ghatsila). Mining areas of these minerals have generally showed higher concentration of SO_4^{2-}, e.g., Govindpur, Maubhandar, Ghatsila, Kandra and Saraikela in groundwater. Overall, groundwater is more polluted than surface water of the SRB.

TABLE 3 The Statistics of the physico-chemical parameters in the surface and groundwater samples.

Variables	Surface water				Groundwater			
	Min	Max	Average	Stdev	Min	Max	Average	Stdev
pH	7.07	7.87	7.61	0.22	5.97	7.67	7.21	0.405
EC (μScm^{-1})	74.33	289.67	207.33	62.14	304.00	2833.33	965.19	578.2
TDS (mgL^{-1})	56.67	238.33	170.95	50.65	225.67	2319.33	805.53	474.11
F^- (mgL^{-1})	0.24	0.92	0.51	0.23	0.12	1.50	0.56	0.368
Cl^- (mgL^{-1})	5.2	29.8	13.53	6.41	20.07	390.37	145.49	109.71
HCO_3^- (mgL^{-1})	27.53	111.5	87.98	24.44	111.27	443.03	254.79	80.59
SO_4^{2-} (mgL^{-1})	2.73	28.13	11.62	8.92	2.80	631.37	109.49	148.49
NO_3^- (mgL^{-1})	1.17	7.63	3.69	2.3	0.27	223.97	46.80	54.40
Silica (mgL^{-1})	5.43	22.73	15.17	4.1	12.87	53.00	38.58	9.062
Ca^{2+} (mgL^{-1})	6.77	27.17	19.29	6.19	23.60	376.07	133.16	86.71
Mg^{2+} (mgL^{-1})	2.4	9.3	6.33	2.06	5.10	139.97	34.25	30.89
Na^+ (mgL^{-1})	3.23	17.37	10.05	4.04	10.07	105.33	39.21	28.047
K^+ (mgL^{-1})	1.17	5.03	2.84	1.3	1.17	12.53	3.21	2.5870

Irrigation Indices

Table 4 shows the overall information of surface and groundwater samples.

Alkali and Salinity Hazard (SAR)

The combined effect of EC and SAR on plant growth is shown graphically by the United States Salinity Laboratory (USSL; Richards 1954; Gautam et al. 2015), which is widely used for classification of water quality for irrigation. EC and sodium concentration are very important in classifying irrigation water. The total concentration of soluble salts in irrigation water can be expressed for the purpose of classification of irrigation water as excellent (EC = <250 μS/cm), good (250–750 μS/cm), permissible (750–2,000 μS/cm), doubtful (2,000–3,000 μS/cm) and unsuitable (> 3,000 μS/cm) on salinity zones (Richards 1954).

The exchangeable sodium percentage/capacity of water is expressed by sodium absorption ratio (SAR) (eq. 3).

$$SAR = \frac{Na^+}{\sqrt{\left(\dfrac{Ca^{2+} + Mg^{2+}}{2}\right)}} \tag{3}$$

where, the cationic concentrations are in milliequivalents per litre (meqL^{-1}).

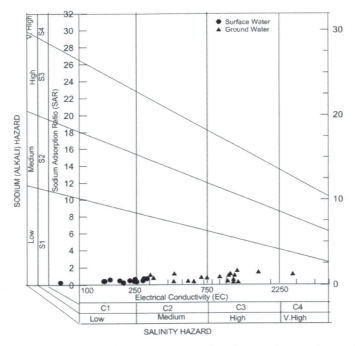

Figure 4 Irrigation water quality assessments of surface and groundwater based on United States Salinity Laboratory (USSL; Richards 1954).

High sodium content in water may have harmful effects in most soils and requires special management of water and soil, which includes application of gypsum. High concentration of bicarbonates and relatively low calcium in water is also known to be hazardous for irrigation (Richards 1954). Salt content in irrigation water causes an increase in soil solution osmotic pressure (Thorne and Peterson 1954). The average calculated value of SAR in the surface and groundwater is 0.5 and 0.82 respectively. The plot of the US salinity diagram has shown that the surface water in the category C1S1 (76.92%) and remaining C2S1 (23.08%) region, indicating low to medium salinity and low alkalinity and groundwater samples fall in category C2S1 (42.10%), C3S1 (52.63%) and remaining 5.26% in C4S1 region, indicating medium to very high salinity and low alkalinity (Fig. 4). The water in the sampling area is more suitable to salt tolerant plants. High saline water cannot be used on soils with restricted drainage and requires special management for salinity control. Low sodium (alkali) water can be used for irrigation on almost all soils. Medium sodium water can create an appreciable sodium hazard in fine textured soils having high cation exchange capacity (CEC) especially under low leaching conditions. The water from the study area can be used on coarse textured or organic soils with good permeability (Karanth 1989).

Sodium Percentage (Na%)

Excess sodium in water produces the undesirable effects of changing soil properties and reducing soil permeability (Kelley 1951; Gautam et al. 2013). The Wilcox (1955) diagram uses sodium percentage, electrical conductivity (Fig. 5) to classify surface and groundwater. Na% is calculated using the formula (eq. 4), where all the concentrations in $meqL^{-1}$.

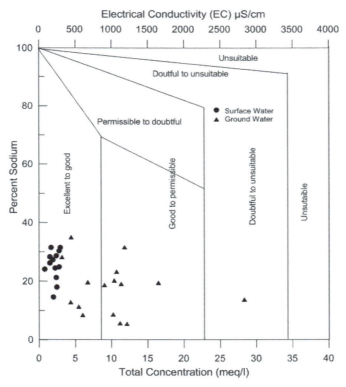

Figure 5 The (Wilcox 1955) based classification of surface and groundwater.

$$Na\% = \frac{Na^+}{Ca^{2+} + Mg^{2+} + Na^+ + K^+} \times 100 \tag{4}$$

Generally, the Na% should not exceed 60% in water as used for irrigation. In the present study, the Na% in surface water samples ranges from less than 20% (2 samples) and 20-40% (11 samples) and in groundwater samples it ranges from less than 20% (13 samples) and 20-40% (6 samples). Bases on the Na% of surface water samples on the Wilcox diagram, samples fall in the categories of excellent (15.38%) to good (84.61%) while for groundwater samples fall in the categories of excellent (68.42%) to good (31.57%). The maximum water samples on the Wilcox diagram fall in the categories of excellent to good for irrigation purposes. According to Wilcox's (1955) classification, the water of SRB is generally suitable for irrigation purpose.

Residual Sodium Carbonate (RSC)

The quantity of bicarbonate and carbonate in excess of alkaline earths ($Ca^{2+} + Mg^{2+}$) also influence the suitability of water for irrigation purposes. When the sum of carbonates and bicarbonates is in excess of calcium and magnesium, there may be a possibility of complete precipitation of Ca^{2+} and Mg^{2+} (Raghunath 1987; Maharana et al. 2015). To quantify the effects of carbonate and bicarbonate, RSC has been computed by the (eq. 5):

$$RSC = (CO_3^- + HCO_3^-) - (Ca^{2+} + Mg^{2+})$$ (5)

A high value of RSC in water leads to an increase in the adsorption of sodium in soil (Eaton 1950). Waters having RSC values greater than 5 $meqL^{-1}$ are considered to be harmful to the growth of plants, while waters with RSC values above 2.5 $meqL^{-1}$ are not considered as suitable for irrigation purposes.

In the present study, the river water samples show RSC values ranging from 0.18–0.92 $meqL^{-1}$ and in groundwater ranges from 7.95-2.54 $meqL^{-1}$. As per RSC classification all the surface water is in the safe region (RSC < 1.25 $meqL^{-1}$) and maximum (78.95%) groundwater samples lie in the safe region and the remaining 21.05% are marginally safe (RSC ranges from 1.25-2.5 $meqL^{-1}$). This indicates that the water is suitable for irrigation uses.

Permeability Index (PI)

Doneen (1964) has classified irrigation waters based on the permeability index (PI). PI is defined by (eq. 6).

$$PI = \frac{(Na^+ + \sqrt{HCO_3^-})}{(Ca^{2+} + Mg^{2+} + Na^+)} \times 100$$ (6)

The soil permeability is affected by the long term use of irrigation activities as it is influenced by the Na^+, Ca^{2+}, Mg^{2+} and HCO_3^- content of the soil (Ishaku et al. 2011). PI of surface water ranges from 69.5-120.2 and in groundwater its values ranges from 18.4-87.6. PI value of surface water samples (Fig. 6), showed that 84.61% samples falls in Class-I (PI > 75, safe) and 15.38% fall in Class-II (PI = 25-75, marginally safe). The PI value of groundwater samples showed that 15.79% groundwater samples fall in Class-I (PI > 75, safe), 68.42% in Class-II (PI = 25-75, marginally safe) and remaining 15.78% in Class-III (PI < 25, unsafe) in the Doneen's chart (Domenico and Schwartz 1990).

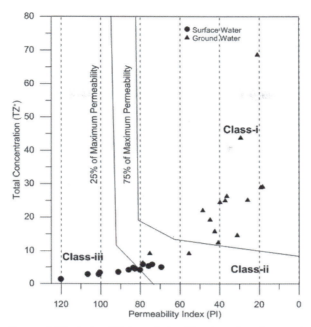

Figure 6 The Permeability Index (PI) based classification of surface and groundwater.

Magnesium Hazard (MH)

Higher concentrations of magnesium in water affect the soil quality by converting it to alkaline and decreases crop yield (Gowd 2005; Singh et al. 2013b). Szabolcs and Darab (1964) proposed magnesium hazard (MH) or magnesium ratio (MR) value for irrigation water as given below by (eq. 7):

$$MH = \frac{Mg^{2+}}{(Ca^{2+} + Mg^{2+})} \times 100 \qquad (7)$$

MH < 50 is considered safe and > 50 is unsafe for irrigation use. In the analyzed water samples, all the samples lie in the safe region (MH values < 50), except at one location of the groundwater sample, it has slightly high MH values (54.2). So that maximum water samples are suitable for irrigation uses.

Kelly's Index (KI)

Kelly's index is used for the classification of water for irrigation purposes (Kelly 1957). Sodium measured against $Ca^{2+} + Mg^{2+}$ is considered to calculate the Kelly index by (eq. 8).

$$KI = \frac{Na^+}{(Ca^{2+} + Mg^{2+})} \qquad (8)$$

where, all the ions are expressed in meq/L^{-1}.

If KI values are greater than 1 (KI > 1), it indicates an excess level of sodium in waters, which is unsuitable for irrigation and KI value less than 1 (KI < 1) is suitable for irrigation purposes. All water samples have KI value < 1 (average 0.24). As per KI values, surface and groundwater of SRB is suitable for irrigation purposes.

TABLE 4 Classification of water quality based on suitability of water for irrigation purposes.

Parameters	Range	Class	No. of surface water samples	% of samples	No. of groundwater samples	% of samples
EC	<250	Excellent	10(1S, 2S, 3S, 5S, 8S, 9S, 10S, 11S, 12S and 13S)	79.92	Nil	Nil
	250-750	Good	3(4S, 6S and 7S)	23.07	8(15G, 17G, 18G, 22G, 26G, 27G, 31G and 32G)	42.10
	750-2,000	Permissible	Nil	Nil	10(14G, 16G, 20G, 21G, 23G, 24G, 25G, 28G, 29G and 30G)	52.63
	2,000-3,000	Doubtful	Nil	Nil	1(19G)	5.26
	>3,000	Unsuitable	Nil	Nil	Nil	Nil
Na%	<20	Excellent	2(10S and 11S)	15.38	13(16G, 17G, 18G, 19G, 20G, 22G, 23G, 25G, 27G, 28G, 29G, 30G and 31G)	68.42
	20-40	Good	11(1S, 2S, 3S, 4S, 5S, 6S, 7S, 8S, 9S, 12S and 13S)	84.61	6(14G, 15G, 21G, 24G, 26G and 32G)	31.57
	40-60	Permissible	Nil	Nil	Nil	Nil
	60-80	Doubtful	Nil	Nil	Nil	Nil
	>80	Unsuitable	Nil	Nil	Nil	Nil

Table 4 Contd.

Parameters	Range	Class	No. of surface water samples	% of samples	No. of groundwater samples	% of samples
RSC	<1.25	Safe	13(1S, 2S, 3S, 4S, 5S, 6S, 7S, 8S, 9S, 10S, 11S, 12S and 13S)	100	15(14G, 16G, 17G, 18G, 19G, 20G, 21G, 22G, 23G, 24G, 28G, 29G, 30G, 31G and 32G)	78.95
	1.25-2.5	Marginally safe	Nil	Nil	4(15G, 25G, 26G and 27G)	21.05
	>2.5	Unsuitable	Nil	Nil	Nil	Nil
PI	>75	Safe	11(1S, 2S, 3S, 5S, 6S, 7S, 8S, 9S, 11S, 12S and 13S)	84.61	3(15G, 26G and 32G)	15.79
	25-75	Marginally safe	2(4S and 10S)	15.38	13(14G, 16G, 17G, 18G, 20G, 21G, 22G, 24G, 25G, 27G, 29G, 30G and 31G)	68.42
	<25	Unsafe	Nil	Nil	3(19G, 23G and 28G)	15.78
MH	<50	Safe	13(1S, 2S, 3S, 4S, 5S, 6S, 7S, 8S, 9S, 10S, 11S, 12S and 13S)	100	18(14G, 15G, 16G, 17G, 18G, 19G, 20G, 21G, 22G, 23G, 24G, 26G, 27G, 28G, 29G, 30G, 31G and 32G)	94.73
	>50	Unsafe	Nil	Nil	1(25G)	5.26
KI	<1	Suitable	13(1S, 2S, 3S, 4S, 5S, 6S, 7S, 8S, 9S, 10S, 11S, 12S and 13S)	100	19(14G, 15G, 16G, 17G, 18G, 19G, 20G, 21G, 22G, 23G, 24G, 25G, 26G, 27G, 28G, 29G, 30G, 31G and 32G)	100
	>1	Unsuitable	Nil	Nil	Nil	Nil

Graphical Representation of Water Quality Data

Gibbs Diagram

The Gibbs (1970) diagram is widely used to establish the relationship of water composition and aquifer lithological characteristics. Three distinct fields such as precipitation dominance, evaporation dominance and rock-water interaction dominance areas are shown in the Gibbs diagram (Fig. 7). The distribution of the samples in the rock dominance region of the plot suggests that the major ion chemistry of the surface and groundwater seems to be controlled by chemical weathering of rock forming minerals and anthropogenic activities.

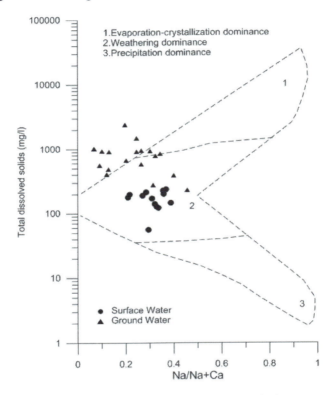

Figure 7 Gibbs diagrams shows water-rock/soil interaction which is responsible for the chemical composition of the surface and groundwater in the region.

Piper Diagram

The Piper's trilinear diagram (Piper 1944) includes two triangles, one for plotting cations and the other for plotting anions (Figure 8). The cations and anion fields are combined to show a single point in a diamond shaped field and inference is drawn for the hydrogeochemical facies of water. Back (1961) and Back and Hanshaw (1965) defined the subdivisions of the diamond field that represent water-type categories that form the basis for one common classification scheme for natural waters. Lithology, solution kinetics and flow patterns of the aquifer control hydrochemistry of the facies. All the samples of surface water are Ca–HCO$_3$ type when plotted on the Piper diagram. But 47.37% groundwater samples fall in Ca-Cl type, 36.84% fall in Ca-HCO$_3$ type, 10.53% fall in Ca-SO$_4$ type and remaining 5.26% groundwater samples fall in Mg-HCO$_3$ type.

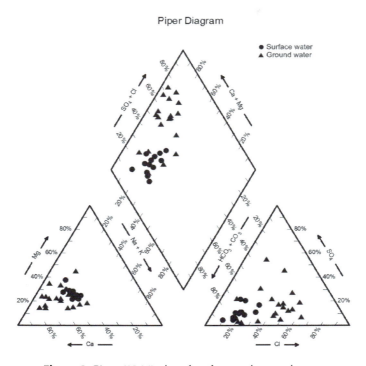

Figure 8 Piper (1944) plot of surface and groundwater.

Durov Diagram

In the Soviet literature, a new diagram was introduced by Durov (1948). An expanded version of the Durov diagram (Fig. 9) was developed by Burdon and Mazloum (1958) and Lloyd (1965). In this diagram, the cations (Na$^+$, K$^+$, Mg^{2+} and Ca^{2+}) and anions (Cl$^-$, SO$_4^{2-}$ and alkalinity or HCO$_3^-$) are plotted in separate triangles, the points are then projected onto a square central field, which is an alternative to the Piper diagram. The Piper diagram is normally based upon the milliequivalent percentage of major-ion values, where total cations and total anions are each considered as 100%, whereas the Durov diagram considers total cations and total anions together as 100%. The Durov plot (Al-Bassam et al. 1997) was also useful in detecting the chemical processes that took place in the aquifer.

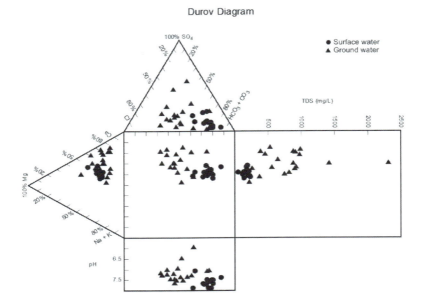

Figure 9 Durov plot of surface and groundwater.

The expanded Durov diagram is used to show the hydro-chemical processes occurring within different hydrogeological systems. The Durov plot for surface and groundwater samples indicates that clusters of samples are in the phase of mixing dissolution. It is also clear from the Durov diagram that the surface and groundwater of the SRB have similar processes in most cases. The pH part of the plot reveals that water samples in the study area are slightly alkaline in nature.

Schoeller Diagram

The Schoeller is semi-logarithmic diagram (Fig. 10) (Schoeller 1956, 1967; Thakur et al. 2013) permits the major ions of many samples to be represented on a single graph, in which samples with similar patterns can be easily discerned (Cetindag and Okan 2003). The Schoeller diagram shows the total concentration of major ions in log-scale, while Schoeller classification (Schoeller 1965) is based on the physical and chemical importance of hydrochemical parameters.

The Schoeller diagram reveals that Ca^{2+} and HCO_3^- are the dominant ions, due to the carbonate dissolution process and wide contact with limestone in surface and groundwater. The geology in the investigated area favors the occurrence of such minerals (Kolhar Limestone and dolomitic limestone). The diagram is also depicting that at a few locations Na^+ cation and Cl^- and SO_4^{2-} anions are dominant in the groundwater samples in the region.

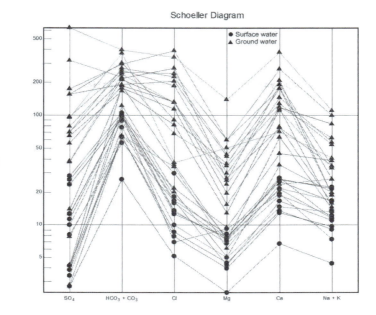

Figure 10 Schoeller plot of surface and groundwater.

Multivariate Statistical Technique

The multivariate statistical analysis has involved cluster analysis (CA). CA has been used in this study to identify the similar groups between the sampling sites. The data sets are treated with hierarchical agglomerative clustering (HAC) following the Wards methods with Euclidean distance for measuring of similarity. The polar cluster diagrams rendered by CA are represented through Fig. 11 (a-b). CA successfully generates the distinct groups. There are five clusters detected for surface water as SW1, SW2, SW3, SW4 and SW5 and similarly for ground water GW1, GW2, GW3, GW4 and GW5. This grouping gives evidence that some sites have similar sources of pollution. The polar diagram reveals that all the stations in GW1 are similar in their behavior; similarly stations in GW2 are purely identified as with matching properties samples. In cluster GW3, many samples share the same properties, indicating that they are influenced by similar types of sources. From LULC it appears that these areas are mostly suffering from urbanized waste and agricultural runoff. In cluster GW4, a mixed response obtained that these stations more or less have similar sources of pollution. In this cluster, many samples are from industrial area, urban as well as agricultural land, so a mixing of contamination cannot be denied. These stations are mainly polluted due to extensive agricultural and industrial activities followed by domestic and municipal discharges. The GW5 also pertains to similar types of parameters. The cluster SW1 has mixed performance. These stations are mainly polluted due to extensive agricultural and industrial activities. The stations in SW2 cluster are mostly from residential areas and hence that's the reason they clustered in the same group with respect to groundwater quality as they belong to the same source of pollution from domestic and municipal discharges. The cluster

3 (i.e. SW3 members) has the same irrigation quality because of similar sources of pollution from runoff. In this group a high variability is obtained as the samples belong to industrial sites, cultivable land, built up, and agricultural land. Because of weathering dominance in the area, a mixing of contaminant cannot be denied. Over here the samples suffer from all sorts of pollution such as industrial waste, agricultural runoff, leaching from waste dumping sites and urban waste. Cluster 4 (SW4) and cluster 5 (SW5) are also linked with similar processes.

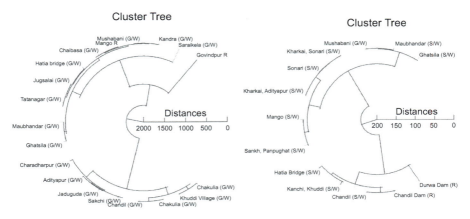

Figure 11 Polar cluster diagram representing similar water quality parameters sites of (a) groundwater (b) surface water.

○ CONCLUSIONS

Surface and groundwater composition is largely controlled by rock weathering and ion exchange processes with secondary contribution from anthropogenic sources. Sodium concentration is important in classifying irrigation water because sodium reacts with soil to reduce its permeability. On the basis of %Na, SAR, RSC, PI, MH and KI values, water of SRB is suitable for irrigation purposes, except at a few locations. Due to mining activities the water quality of the region may have very adverse effects on a long term basis. The distribution of the samples in the rock dominance region of the Gibbs plot suggests that the major ion chemistry of the surface and groundwater seems to be controlled by chemical weathering of rock forming minerals and anthropogenic activities. As per Piper diagram all the samples of surface water are Ca-HCO$_3$ type. But 47.37% groundwater samples fall in Ca-Cl type, 36.84% fall in Ca-HCO$_3$ type, 10.53% fall in Ca-SO$_4$ type and the remaining 5.26% groundwater samples fall in Mg-HCO$_3$ type. Concentration of SO$_4^{2-}$ exceed the desirable limit at a few sites and require treatment before its utilization. The Durov plot for surface and groundwater samples exposes that cluster of samples are in the phase of mixing dissolution in the SRB. The Schoeller diagram reveals that Ca^{2+} and HCO$_3^-$ are the dominant ions, due to the carbonate dissolution process and wide contact with limestone in surface and groundwater. The separate polar cluster diagram of both the surface and ground water in this study suggests that the similar sites forms a group. Total five clusters detected for each surface water as SW1, SW2,

SW3, SW4 and SW5 and similarly for ground water GW1, GW2, GW3, GW4 and GW5. Hence this information will help in monitoring the similar/dissimilar sites in the study area.

○ ACKNOWLEDGMENTS

We are thankful to the reviewers for their valuable suggestions and recommendation. The first author is also thankful to the CSIR, New Delhi, India for providing the financial support. The author also acknowledges the technical and administrative staff of his school for being always attentive and helpful during analysis.

○ REFERENCES

Al-Bassam, A.M., Awad, H.S. and Al-Alawi, J.A. 1997. Durov plot: a computer program for processing and plotting hydrochemical data. Ground Water 35(2): 362-367.

APHA (American Public Health Association) 2005. Standard Methods for the Examination of Water and Wastewater, 21st ed. American Public Health Association, Washington DC.

Back, W. 1961. Hydrochemical facies and groundwater flow patterns in northern part of Atlantic Coastal Plain. US Geological Survey Professional Paper 1966, 498-A. USGS, Washington DC.

Back, W. and Hanshaw, B.B. 1965. Chemical geohydrology. Adv Hydroscience 2: 49-109.

Berner, E.K. and Berner, R.A. 1987. The Global Water Cycle: Geochemistry and Environment. Prentine Hall, Englewood Cliffs, New Jersey.

BIS (Bureau of Indian Standards) IS10500: 2004-05. Specification for Drinking Water, Indian Standard Institution. New Delhi, India.

Bishop, Y.M.M., Fienberg, S.E. and Holland, P.W. 1975. Discrete Multivariate Analysis Theory and Practice. MIT Press, Cambridge, Massachusetts, 557 p.

Burdon, D.J. and Mazloum, S. 1958. Some chemical types of groundwater from Syria. UNESCO Symp, Teheran 73-90.

Cetindag, B. and Okan, O.O. 2003. Hydrochemical characteristics and pollution potential of Uluova aquifers, Elazig, Turkey. Environmental Geology 45: 796-807.

Chidambaram, S., Anandhan, P., Prasanna, M., Srinivasamoorthy, K. and Vasanthavigar, M. 2013. Major ion chemistry and identification of hydrogeochemical processes controlling groundwater in and around Neyveli Lignite Mines, Tamil Nadu, South India. Arabian Journal of Geosciences 6(9): 3451-3467.

Domenico, P.A. and Schwartz, F.W. 1990. Physical and Chemical Hydrogeology, Vol. 824. Wiley, New York.

Doneen, L.D. 1964. Notes on water quality in agriculture. Water science and engineering paper 4001, Department of Water Sciences and Engineering, University of California, California.

Durov, S.A. 1948. Natural waters and graphic representation of their compositions. Dokl Akad Nauk SSSR 59: 87-90.

Eaton, F.M. 1950. Significance of carbonates in irrigation waters. Soil Science 69(2): 123-134.

Foster, S., Hirata, R., Gomes, D., D'Elia, M. and Paris, M. 2002. Groundwater Quality Protection: A Guide for Water Utilities, Municipal Authorities and Environment Agencies. World Bank, Washington, DC.

Gautam, S.K., Sharma, D. Tripathi, J.K., Ahirwar, S. and Singh, S.K. 2013. A study of the effectiveness of sewage treatment plants in Delhi region. Applied Water Science 3: 57–65. DOI 10.1007/s13201-012-0059-9.

Gautam, S.K., Maharana, C., Sharma, D., Singh, A.K., Tripathi, J.K. and Singh, S.K. 2015. Evaluation of groundwater quality in the Chotanagpur Plateau region of the

Subarnarekha River Basin, Jharkhand State, India. Sustainability of Water Quality and Ecology. doi:10.1016/j.swaqe.2015.06.001.

Ghosh, S.K, Sengupta, S. and Dasgupta, S., 2002. Tectonic deformation of soft-sediment convolute folds. Journal of Structural Geology 24: 913-923.

Gibbs, R.J. 1970. Mechanism controlling world water chemistry. Science 170: 795-840.

Gowd, S.S. 2005. Assessment of groundwater quality for drinking and irrigation purposes: a case study of Peddavanka watershed, Anantapur District, Andhra Pradesh, India. Environmental Geology 48(6): 702-712.

Gupta, M. and Srivastava, P.K. 2010. Integrating GIS and remote sensing for identification of groundwater potential zones in the hilly terrain of Pavagarh, Gujarat, India. Water International 35(2): 233-245.

Ishaku, I., Ahmed, A. and Abubakar, M. 2011. Assessment of groundwater quality using chemical indices and GIS mapping in Jada area, Northwestern Nigeria. Journal of Earth Sciences and Geotechnical Engineering 1(1): 35-60.

Jha, P.K., Tiwari, J., Singh, U.K. Kumar, M. and Subramanian, V. 2009. Chemical weathering and associated CO_2 consumption in the Godavari river basin, India. Chemical Geology 264: 364-374.

Karanth, K.R. 1989. Hydrogeology. Tata McGraw-Hill, New Delhi.

Kelley, K.R. 1951. Alkali Soils their Formation Properties and Reclamation. Reinold Publ. Corp., New York.

Kelly, W.P. 1957. Adsorbed sodium cation exchange capacity and percentage sodium sorption in alkali soils. Science 84: 473-477.

Krishna, A.K., Satyanarayanan, M. and Govil, P.K. 2009. Assessment of heavy metal pollution in water using multivariate statistical techniques in an industrial area: a case study from Patancheru, Medak District, Andhra Pradesh, India. Journal of Hazardous Materials 167(1): 366-373.

Lloyd, J.W. 1965. The hydrochemistry of the aquifers of north-eastern Jordan. Journal of Hydrology 3: 319-330.

Maharana, C., Gautam, S.K., Singh, A.K. and Tripathi, J.K. 2015. Major ion chemistry of the Son River, India: Weathering processes, dissolved fluxes and water quality assessment. Journal of Earth System Science 124(6): 1293-1309.

Merchant, J.W. 1994. GIS-based groundwater pollution hazard assessment: a critical review of the DRASTIC model. Photogrammetric Engineering and Remote Sensing 60(9): 1117-1128.

Mukhopadhyay, S.C. 1980. Geomorphology of Subarnarekha Basin, Univ. of Burdwan, Burdwan.

Négrel Ph., Lemière, B., Grammont, H., Machard, de., Billaud, P. and Sengupta, B. 2007. Hydrogeochemical processes, mixing and isotope tracing in hard rock aquifers and surface waters from the Subarnarekha River Basin, (East Singhbhum District, Jharkhand State, India). Hydrogeology Journal 15: 1535-1552.

Piper, A.M. 1944. A graphic procedure in the chemical interpretation of water analysis. Transactions – American Geophysical Union 25: 914-923.

Raghunath, H.M. 1987. Groundwater. Wiley Eastern Ltd., Delhi.

Ramakrishnan, M. and Vaidyanadhan, R. 2008. Geology of India. Geological Society of India, Bangalore 1: 210-213.

Rao, K.L., 1995. India's Water Wealth: Its Assessment, Uses and Projections. Longman, New Delhi. Reprinted.

Richards, L.A. 1954. Diagnosis and Improvement of Saline and Alkali Soils. US Department of Agriculture Handbook 60.

Schoeller, H. 1956. Hydrodynamique lans lekarst (ecoulemented emmagusinement). Actes Colloques Doubronik, I, AIHS et. UNESCO, 3-20.

Schoeller, H. 1965. Qualitative evaluation of groundwater resource. pp. 54-83. *In*: Methods and Techniques of Ground-Water Investigation and Development. UNESCO.

Schoeller, H. 1967. Geochemistry of groundwater. An international guide for research and practice. UNESCO 15: 1-18.

Singh, K.P., Malik, A. and Sinha, S. 2005. Water quality assessment and apportionment of pollution sources of Gomti river (India) using multivariate statistical techniques—A case study. Analytica Chimica Acta 538(1): 355-374.

Singh, S.K., Singh, C.K., Kumar, K.S., Gupta, R. and Mukherjee, S. 2009. Spatial temporal monitoring of groundwater using multivariate statistical techniques in Bareilly district of Uttar Pradesh, India. Journal of Hydrology and Hydromechanics 57(1): 45-54.

Singh, S.K., Srivastava, P.K., Gupta, M. and Mukherjee, S. 2012. Modeling mineral phase change chemistry of groundwater in a rural-urban fringe. Water Science and Technology 66(7): 1502-1510.

Singh, S.K., Srivastava, P.K. and Pandey, A.C. 2013a. Fluoride contamination mapping of groundwater in northern India integrated with geochemical indicators and GIS. Water Science and Technology 13(6): 1513-1523. doi:10.2166/ws.2013.160.

Singh, S.K., Srivastava, P.K., Pandey, A.C. and Gautam, S.K. 2013b. Integrated assessment of groundwater influenced by a confluence river system: concurrence with remote sensing and geochemical modelling. Water Resources Management 27(12): 4291-4313.

Singh, S.K., Srivastava, P.K., Singh, D., Han, D., Gautam, S.K. and Pandey, A.C. 2015. Modelling groundwater quality over a humid subtropical region using numerical indices, earth observation datasets and X-ray diffraction technique: a case study of Allahabad district, India. Environmental Geochemistry and Health 37: 157-180. DOI 10.1007/s10653-014-9638-z.

Srivastava, P.K., Han, D., Gupta, M. and Mukherjee, S. 2012b. Integrated framework for monitoring groundwater pollution using a geographical information system and multivariate analysis. Hydrological Sciences Journal 57(7): 1453-1472. doi:10.1080/02626667.2012.716156.

Srivastava, P.K., Han, D., Rico-Ramirez, M.A., Bray, M. and Islam, T. 2012a. Selection of classification techniques for land use/land cover change investigation. Advances in Space Research 50(9): 1250-1265. doi:10.1016/j.asr.2012.06.032.

Srivastava, P.K., Singh, S.K., Gupta, M., Thakur, J.K. and Mukherjee, S. 2013. Modeling impact of land use change trajectories on groundwater quality using remote sensing and GIS. Environmental Engineering and Management Journal 12(12): 2343-2355.

Sun, H., Bergstrom, J.C. and Dorfman, J.H. 1992. Estimating the benefits of groundwater contamination control. Southern Journal of Agricultural Economics 24: 63.

Szabolcs, I. and Darab, C. 1964. The influence of irrigation water of high sodium carbonate content of soils. *In*: Proceedings of 8th International Congress of Isss, Trans, vol II: 803-812.

Thakur, J.K., Singh, P., Singh, S.K. and Bhaghel, B. 2013. Geochemical modelling of fluoride concentration in hard rock terrain of Madhya Pradesh. India Acta Geologica Sinica (English Edition) 87(5): 1421-1433.

Thorne, D.K. and Peterson, H.B. 1954. Irrigated Soils. Constable and Company, London.

Vasanthavigar, M., Srinivasamoorthy, K., Vijayaragavan, K., Ganthi, Rajiv, R., Chidambaram, S., Anandhan, P., Manivannan, R. and Vasudevan, S. 2010. Application of water quality index for groundwater quality assessment: Thirumanimuttar sub-basin, Tamilnadu, India. Environmental Monitoring and Assessment 171(1-4): 595-609.

Waller, R.M. 2001. Ground Water and the Rural Homeowner. US Geological Survey (USGS).

Wilcox, L.V. 1955. Classification and use of irrigation waters, USDA Circular No. 969, 19.

Wu, J. and Segerson, K. 1995. The impact of policies and land characteristics on potential groundwater pollution in Wisconsin. American Journal of Agricultural Economics 77(4): 1033-1047.

11

CHAPTER

◇◇

Spatial and Temporal Variability of Sea Surface Height Anomaly and its Relationship with Satellite Derived Chlorophyll *a* Pigment Concentration in the Bay of Bengal

Preeti Rani * and *S. Prasanna Kumar*

ABSTRACT

This study focuses on the spatial and temporal variability of sea surface height anomaly (SSHA) using satellite altimeter data sets. The data used are the monthly maps of sea level height anomaly with spatial resolution of $1/3°$ latitude and $1/3°$ longitude for the period January 1997 to December 2010. Annual average of sea surface height anomaly field is estimated to understand the long term variability. The data on (SSHA) derived from satellite altimeter and chlorophyll *a* pigment concentration derived from ocean color were subjected to Empirical Orthogonal Function (EOF) analysis to understand the Spatio-temporal variability of these parameters in the context of mesoscale eddies. Empirical Orthogonal Function (EOF) is an efficient method of delineating a spatial and temporal signal from a long time series data over a large spatial domain such as basin. EOF has been carried out to derive the spatial structure and temporal amplitude of the variability and described the annual, semi-seasonal and inter-annual SSHA variations. EOFs of SSHA showed that annual cycle accounts for 54% of the total variance by first 2 EOF modes and remaining 56% of variance showed, semi-annual, intra-annual and mesoscale eddies variability but chlorophyll *a* pigment concentration

Department of Physical Oceanography, National Institute of Oceanography, Dona Paula, North Goa, India.

* Corresponding author: preetiprachi2007@gmail.com

accounted only 25% of the total variance for the annual periodicity. The spatial patterns of EOF modes resembled the seasonal cycle of circulation near the western boundary of the Bay of Bengal. In contrast, spatial EOFs of chlorophyll showed only the dominance of annual variability.

This study shows a disagreement between the spatial EOF of SSHA and chlorophyll *a* concentration and points to the fact that in the Bay of Bengal the process controlling the chlorophyll enhancement is not the traditional upwelling but also the mesoscale variability associated with eddies and nutrient input from river discharge.

KEYWORDS: Bay of Bengal, Sea Level Height Anomaly (SSHA), Chlorophyll *a* pigment concentration, Empirical orthogonal function, Variability.

○ INTRODUCTION

The Bay of Bengal, located in the western part of the north Indian Ocean is a tropical basin. It is land locked in the north at around 30°N and to the south it is open to the Indian Ocean. To the west it is bounded by the Indian land mass, while to the east it is bounded by the East Asian land mass. It is only the southern part of the Bay of Bengal that is open to the Indian Ocean. During summer, the heating of Asian land mass generates large land-sea pressure difference, which drives the southwesterly (summer) Monsoon winds. In contrast, during winter the Asian land mass is much colder than the ocean which drives the northeasterly (winter) monsoon winds. The southwesterly winds are strong (10 m/s), while the northeasterly winds are weak (5 m/s). During summer the moist southwesterly monsoon winds bring a large amount of precipitation to the Bay of Bengal. The surface currents in the Bay of Bengal also show a seasonal reversal. This reversal in the surface current is most prominently seen in the coast currents along the east coast of India. The East India Coastal Current (EICC) moves southward during the winter monsoon transporting warm low salinity waters from the head of Bay of Bengal towards the equator. The same current changes its direction during the summer monsoon transporting high salinity waters of the Arabian Sea in the northward direction. In addition to the reversal of the coastal currents, the currents in the open Bay of Bengal also change their direction. For example, the north equatorial current in the Bay of Bengal moves westward in winter while in summer it moves from west to the east. (Wyrtki 1971) presents initial information of Bay of Bengal waters and (Prasanna Kumar et al. 1993) described the seasonal and inter-annual sea surface height variations using TOPEX/POSEIDON altimeter data. [Varkey et al., 1996; Singh et al., 2015] shows the seasonal variation and circulation in the Bay of Bengal. Recent studies have been examined the surface circulation and mesoscale features in the Bay of Bengal and (Benny and Mizuno 2000) described the Kelvin and Rossby waves in the Bay of Bengal. Life cycle of eddies along the western boundary of the Bay of Bengal and their implication is well explained by Nuncio et al. in (2012).

Satellite altimetry derived sea surface height anomaly data have produced a large scale circulation feature observed in all the major oceans such as Pacific Ocean, Atlantic Ocean, Indian Ocean, Arctic Ocean and Antarctic/Southern Ocean. They are highly energetic motions of water and exhibit great variations both in time and space. The variability in the ocean currents arises due to a host of features like rings, vortices, lens, meanders, jets and filaments. Physical oceanographers group them in a generic term 'eddies' (Robinson 1983). The present study examines the variability of the monthly sea surface height anomaly of Bay of Bengal during January 1997 to December 2010 using satellite altimetry observations and examines the spatial structures and temporal variability of the Bay of Bengal with the help of Empirical Orthogonal Function.

○ DATA AND METHODOLOGY

Satellite altimetry uses microwave remote sensing. It measures sea surface height by measuring the time taken by a radar (microwave) pulse to travel from the satellite antenna to the surface of the ocean and back to the satellite receiver. The time measurement scaled by the speed of light in (electromagnetic wave) yields a range measurement. Combined with precise satellite location data, altimetry measurements yield sea-surface heights. Thus, by measuring the sea surface height variations in a given oceanic region one could study about the ocean circulation.

In the present study, the 7-day snapshots of the merged sea-surface height anomalies from Topex-Poseidon/ERS1/2/Jason1/2 satellites were obtained from AVISO live access server (http://las.aviso.oceanobs.com). This data, having a spatial resolution of 1/3rd of a degree, were extracted for the period from January 1997 to December 2010 (Le Traon et al. 1998), which was used to infer about the mesoscale variability.

In addition to the above weekly data, the monthly mean sea surface height anomaly (SSHA) data was extracted from the Asian Pacific data research centre (APDRC), which contains AVISO TOPEX/ERS/Jason merged gridded data having a spatial resolution of 1/3° latitude by 1/3° longitude (http://apdrc.soest.hawaii.edu/las/v6/constrain?var=2563) for the period January 1997 to February 2010.

In ocean color remote sensing, the color of the ocean is determined by the interactions of visible part of the incident light with substances or particles present in the water. In the ocean blue wavelengths of light are scattered while the red wavelengths are absorbed strongly. In the open ocean important substances that influence the color of the ocean are phytoplankton, suspended material and yellow substances. Hence, optical sensors onboard satellite can quantify the chlorophyll present in the ocean by studying the color of the light scattered from the oceans.

The monthly mean chlorophyll *a* pigment concentration data was obtained from ocean color radiometry. For the present study, two satellite data products (1) SeaWiFS (Sea-viewing Wide Field-of-view Sensor) and (2) MODIS (The Moderate-resolution Imaging Spectro-radiometer) (Tomar et al., 2014) Aqua was used. The SeaWiFS data has a spatial resolution of 9-km and though available from September 1997, for the present study data for the period January 1998 to December 2010 was used. While analyzing the data it was found that the data for the months February and March 2008 and May 2009 were not available globally. In addition, the data for the Bay of Bengal region was missing during January and July 2008 while September and October in 2009. Hence, for the missing data, the chlorophyll *a* pigment concentration from

MODIS Aqua satellite having the same spatial resolution was used. Both the data were downloaded from the link (http://oceancolor.gsfc.nasa.gov/).

To understand the spatial and temporal variations of sea level height anomalies in the Bay of Bengal-empirical Orthogonal Function (EOF) analysis is found efficient, as it is capable of conveying most of the original data without redundancy and with a relatively small number of independent variables (factors). EOF analysis can be used to explore the structure of the variability within a data set in a objective way and to analyze relationships within a set of variables. EOF analysis is also called principal component analysis or factor analysis. In brief, EOF analysis uses a set of orthogonal functions (EOFs) to represent a time series in the following way.

$$Z(x, y, t) = \sum_{k=1}^{N} PC(t), EOF(x, y)$$

whereas, $Z(x, y, t)$ is the original time series as a function of time (t) and space (x, y). EOF (x, y) shows the spatial structures (x, y) of the major factors that can account for the temporal variations of Z. $PC(t)$ is the principal component that tells you how the amplitude of each EOF varies with time. The first four modes of EOF explain the highest division of the variance and the remaining percentage of variances computed in other modes.

○ RESULTS

The present study describes the annual average of monthly SSHA from January 1997 to December 2010. The long term variability and change in climate mode is determined from the first and first four EOFs and corresponding time function.

Annual Average of Sea Surface Height Anomaly (SSHA)

The annual average of monthly SSHA patterns shows much variation in the Bay of Bengal during January 1997 to December 2010. The salient feature of the annual average SSHA was the presence of both positive and negative height anomaly in the Bay of Bengal, especially along the western boundary (as shown in Fig. 1). During 1997 a positive height anomaly was seen predominantly along the northwestern boundary, close to the coast at about 18°N, 86°E. Meanwhile another positive height anomaly appears in the south, northwest of Indo-Sri Lanka and central Bay of Bengal and a negative anomaly shown along the 14°N, 80-84°E in annual average of 1997. In 1998, the SSHA showed negative height anomaly near the indo-Sri Lanka along the central Bay of Bengal and rest of the Bay exhibits positive values. But, in 1999 a decrease in SSHA occurred in the Bay except in the northeast part. SSHA slightly increased the northward side in 2000 and 2001. The central and southern part of Bay of Bengal showed high SSHA in 2002 and its increase towards the north side. A complete portion of Bay of Bengal showed a very high value of height anomaly except a small portion along the north and northwest boundary during 2003 and 2007. The central-eastern part of Bay of Bengal showed high values in 2004 as well as its slight increase towards north-west in 2005. During 2006, SSHA showed the high value along southern and central-western boundary near Sri-Lanka and its slightly increase towards the north. The southern part and central-eastern boundary of Bay of Bengal showed high positive values in 2008. The result for the year 2009-2010 showed almost normal low and high values of height anomaly.

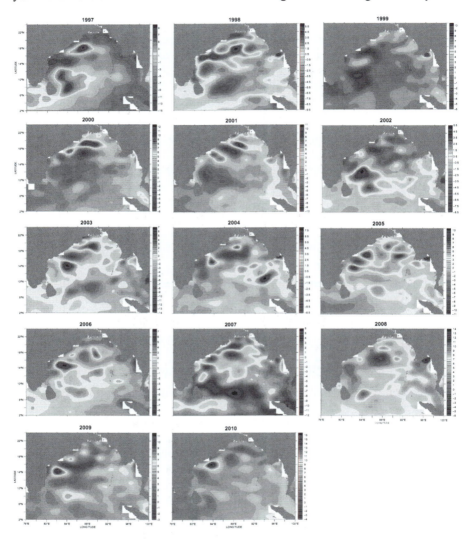

Figure 1 Monthly mean climatology of Topex/Poseidon ERS1/2 merged sea surface height anomalies overlaid with geostrophic velocities during January 1997 to December 2010.

Spatial and Temporal Variability

The empirical orthogonal functional analysis has been shown to be a useful method to extract spatial and temporal variability that are orthogonal in nature from sea surface height anomaly data in the Bay of Bengal (Nuncio and Kumar 2012). They described annual, semi-annual and intra-seasonal variability of the Bay of Bengal and showed that the first three modes of EOFs capture 64% of the total variance of the SSHA in the Bay of Bengal. The intra-seasonal variability was associated with mesoscale eddies whose life cycle varied from 3 to 5 months. The EOF showed that these mesoscale features accounted for about 20% of the total variance (shown in Fig. 2).

In the present case, the first spatial EOF (EOF-1) of the SSHA represents a large-scale pattern with positive (increase) SSHA along the eastern boundary and northern part of the Bay of Bengal. The rest of the Bay exhibits negative (decrease) SSH anomaly and explains about 42% of the total variance. A similar pattern of SSHA is discernible in annual average of SSHA during 1997 with an opposite polarity along the western boundary.

The 2nd spatial EOF of SSHA represents positive values along the western boundary and negative values in the rest of the region. The EOF-2 captures 12% of the total variance and the temporal amplitude of EOF-2 seems to be the composite signal of annual signal of annual cycles. The 3rd spatial function of EOF (EOF-3) explains 10% of the total variance. The 4th EOF explains only 5% of the total variance. The western boundary had the negative SSHA within which there were two distinct regions of very low SSHA. The northern Bay was characterized by strong positive SSHA, the size and shape of which resembled an anticyclonic mesoscale feature. The large band of positive SSHA in the southwest-northeast direction occupied most of the Bay of Bengal. The temporal periodicity of the 4th EOF was about 4 months.

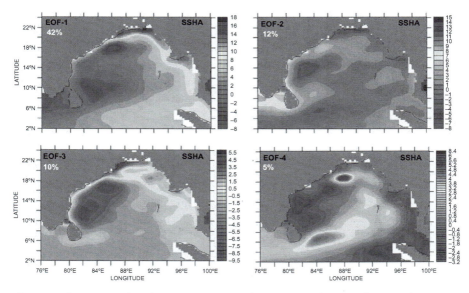

Figure 2 The Empirical Orthogonal Function of the SSHA is showing the spatial structure pattern. The SSHA was from January 1997 to December 2010.

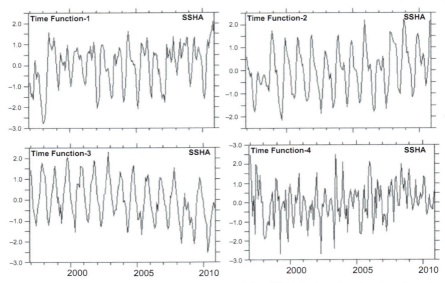

Figure 3 The Empirical Orthogonal Function of the SSHA is showing the time function, where SSHA was from January 1997 to December 2010.

Relationship with Surface Chlorophyll *a* Pigment Concentration

The first spatial EOF mode (EOF-1) explains 25% of variance in the chlorophyll *a* pigment distribution (Fig. 3). The 2nd spatial EOF mode (EOF-2), which showed a positive value along the western boundary explained 6% of the total variance. The third and the fourth spatial EOFs together account for 9% of the total variance.

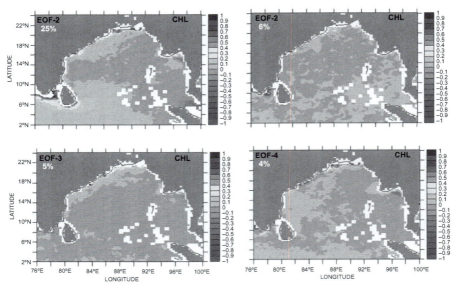

Figure 4 The Empirical Orthogonal Function of the chlorophyll is showing the spatial structure pattern from January 1998 to December 2010.

A comparison of the chlorophyll *a* pigment concentrations with that of SSHA did not show any general correspondence among them indicating that in the Bay of Bengal eddy-induced changes do not significantly influence the surface variability. However, near the Indo-Sri Lankan region a tight correspondance was seen with low SSHA coinciding with high chlorophyll *a* pigment concentration.

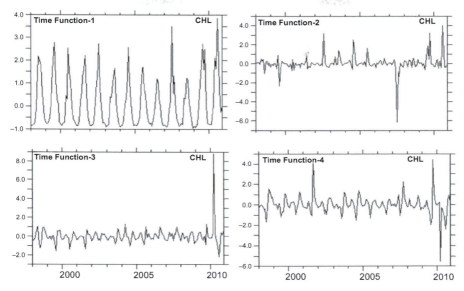

Figure 5 The Empirical Orthogonal Function of the SSHA is showing the time function of chlorophyll from January 1998 to December 2010.

◯ SUMMARY AND CONCLUSION

The SSHA circulation of Bay of Bengal exhibits significant spatial and temporal variations during January 1997 to December 2010. EOF analysis of SSHA reveals that the seasonal circulation (1st, 2nd and 3rd EOFs) accounted for 64% of the total variance while the intra-annual and mesoscale eddies contributed to 36% of the variance, but chlorophyll *a* pigment concentration accounted only for 25% of the total variance for the annual periodicity. The spatial patterns of EOF modes resembled the seasonal cycle circulation near the western boundary of the Bay of Bengal. In contrast, spatial EOFs of chlorophyll showed only the dominance of annual variability. The disagreement between the spatial EOF of SSHA and chlorophyll *a* concentration points to the fact that in the Bay of Bengal process controlling the chlorophyll enhancement is not the traditional upwelling but also the mesoscale variability associated with eddies and nutrient input from river discharge. But the subsurface variability to a large extent could be accounted by the presence of mesoscale eddies. Near the Indo-Sri Lankan region a tight correspondence between low SSHA with high chlorophyll *a* pigment concentration is evident, indicated the role of upwelling, high sedimentation and the fresh water flow from the Brahmaputra-Meghan Rivers in the northern region.

○ REFERENCES

Benny, N.P. and Mizuno, K. 2000. Annual cycle of steric heights in the Indian Ocean estimated from thermal field. Deep-Sea Research, Part I 47: 1351-1368.

Le Traon, P.Y., Nadal, F. and Ducet, N. 1998. An improved mapping method of multisatellite altimeter data. Journal of Atmospheric and Oceanic Technology 15: 522-534.

Nuncio, M. and Prasanna Kumar, S. 2012. Life cycle of eddies along the western boundary of the Bay of Bengal and their implications. Journal of Marine System 94: 9-17.

Prasanna Kumar, S., Snaith, H., Challenor, P. and Guymer, H.T. 1998. Seasonal and interannual sea surface height variations of the northern Indian Ocean from the TOPEX/POSEIDON altimeter. Indian Journal of Marine Science 27: 10-16.

Prasanna Kumar, S., Nuncio, M., Narvekar, J., Kumar, A., Sardesai, S., De Souza, S.N., Gauns, M., Ramaiah, N. and Madhupratap, M. 2004. Are eddies nature's trigger to enhance biological productivity in the Bay of Bengal, Geophysical Research Letter 31: L07309, doi:10..1029/2003G1019274.

Prasanna Kumar, S., Nuncio, M., Ramaiah, N., Sardesai, S., Narvekar, Jayu, Fernandes, Veronica, Paul and Jane, T. 2007. Eddy-mediated biological productivity in the Bay of Bengal during fall and spring inter-monsoons. Deep-Sea Research 54: 1619-1640.

Prasanna Kumar, S., Nuncio, M., Narvekar, J., Ramaiah, N., Sardessai, S., Gauns, M., Fernandes, V., Paul, J.T., Jyothibabu, R. and Jayaraj, K.A. 2010. Seasonal cycle of physical forcing and biological response in the Bay of Bengal. Indian Journal of Marine Science 39: 388-405.

Robinson, A.R. 1983. Eddies in Marine Science. Springer-Verlag, Berlin, Germany, 609pp.

Singh, Kartar, Ghosh, Mili, Sharma, S.R. and Kumar, Pavan. 2014. Blue-Red-NIR Model for Chlorophyll- a Retrieval in Hypersaline-Alkaline Water Using Landsat ETM+ Sensor. IEEE Journal of Selected Topics in Applied Earth Observations and Remote Sensing 7(8): 3553-3559.

Vandana, Tomar, Mandal, Vinay Prasad, Srivastava, Pragati, Patairiya, Shashikanta, Singh, Kartar, Ravisankar, Natesan, Subash, Natraj and Kumar, Pavan. 2014. Rice Equivalent Crop Yield Assessment Using MODIS Sensors' Based MOD13A1-NDVI Data. IEEE Sensors Journal, 13(6): 2161-2165.

Varkey, M.J., Murthy, V.S.N. and Suryanarayana, A. 1996. Physical oceanography of Bay of Bengal and Andaman Sea. Oceanography and Marine Biology 34: 1-70.

Wyrtki, K. 1971. Oceanographic Atlas of the International Indian Ocean Expedition. National Science Foundation, Washington, 531 pp.

12
CHAPTER

✧✧

Monitoring Soil Moisture Deficit Using SMOS Satellite Soil Moisture: Correspondence through Rainfall-runoff Model

Prashant K. Srivastava [1,2,3]

ABSTRACT

Soil Moisture and Ocean Salinity (SMOS) is a dedicated mission, which provides a flow of coarse resolution soil moisture data for hydrological applications. On the other hand Soil Moisture Deficit (SMD) is a key indicator of soil water content changes and is valuable to a wider range of applications such as weather research, climatology, flood, drought, agricultural forecasting etc. In this paper, the SMOS Level 2 soil moisture was used for the estimation of SMD over the Brue catchment, Southwest of England, United Kingdom. Several approaches for estimation of SMD are performed using the Generalized Linear Model with different families/link functions such as Gaussian/logit, Binomial/identity, Gamma/inverse and Poisson/log. The overall performance obtained after all the techniques indicate that the Binomial with identity and Poisson with log link functions look promising for simulation of SMD from SMOS soil moisture with marginally high performance as compared to the other techniques.

KEYWORDS: SMOS, Soil Moisture Deficit, rainfall-runoff model, Generalized Linear Model.

[1] Hydrological Sciences, NASA Goddard Space Flight Center, Greenbelt, Maryland, USA.
[2] Earth System Science Interdisciplinary Center, University of Maryland, Maryland, USA.
[3] Institute of Environment and Sustainable Development, Banaras Hindu University, Varanasi, India.
  E-mail: prashant.just@gmail.com

○ INTRODUCTION

Soil moisture has proven its importance for hydrological sciences and applications (Srivastava 2013). It is a key variable in the water and energy exchanges that occur at the land-surface/atmosphere interface and conditions the evolution of weather and climate over continental regions (Smith 1986; Srivastava et al. 2013a; Srivastava et al. 2014b). The top Earth's surface soil moisture is widely recognised as a key variable in numerous environmental studies including meteorology, hydrology, agriculture and climate change (Jackson 1993; Jackson et al. 1999; Jackson et al. 1995; Mladenova et al. 2011). Therefore, it's timely and accurate monitoring at various spatial and temporal variations are required for improved prediction of environmental variables. Due to enormous applications of soil moisture, one after the other two global space missions have been proposed to provide the global measurements of the Earth's surface soil moisture. The first one that is Soil Moisture and Ocean Salinity (SMOS) mission was launched by the European Space Agency in November 2009, while the second is launched by the National Aeronautics and Space Administration (NASA), that is, Soil Moisture Active and Passive (SMAP) mission in January 2015.

Regardless of the importance of soil moisture information that have an impact on the climate system through atmospheric feedback (Al Bitar et al. 2012; Jackson et al. 2012; Petropoulos et al. 2014; Schlenz et al. 2012; Zhuo et al. 2015), widespread and/or continuous measurements of one important variable i.e. soil moisture deficit (SMD) have been rarely reported in the literature (Srivastava et al. 2013c; Srivastava et al. 2014b). SMD is valuable to a wide range of government agencies and companies concerned with the weather and climate, runoff potential and flood control, drought, soil erosion and slope failure, reservoir management, geotechnical engineering and for crop insurance (Srivastava 2013; Srivastava et al. 2013b; Srivastava et al. 2012). SMD has direct application to hydrological modeling as a soil moisture accounting scheme (Beven and Wood 1983; Croke and Jakeman 2004; Evans and Jakeman 1998) and used by number of agencies in UK such as UK Met office (www.metoffice.gov.uk/), Land Information System (LandIS) (http://www.landis.org.uk/services/seismic.cfm) and Centre for Ecology and Hydrology Wallingford (http://www.ceh.ac.uk/data/nrfa/nhmp/evaporation_smd.html), UK. As a result, accurate information of SMD is needed for improving the discharge prediction capabilities.

In hydrology, soil moisture deficit (SMD) or depletion is a useful soil moisture indicator which is directly related to the ratio between actual and potential evapotranspiration (PE) (Moore 2007; Moran et al. 1994; Narasimhan and Srinivasan 2005). It represents the amount of water required to raise the soil-water content of the crop root zone to field capacity (Calder et al. 1983; Rushton et al. 2006). Moore in (2007) also found that the Probability Distributed Model (PDM) model is very useful in estimating SMD from the hydro-meteorological data. A number studies on PDM model are reported for rainfall-runoff modelling over the globe (Cabus 2008; Liu and Han 2010b; Tripp and Niemann 2008). It could be a suitable option for SMD generation using the minimal datasets such as rainfall, PE and discharge information, unlike the semi distributed models such as the Soil and Water Assessment Tool (SWAT) which requires a number of other ancillary datasets such as land cover, topography and soil databases (Narsimlu et al. 2015; Patel and Srivastava 2013; Patel and Srivastava 2014).

In purview of the importance of SMD, in this study, several attempts have been made to estimate the SMD using the soil moisture information from SMOS by following the different family and link functions available with the generalized linear models. The knowledge gained from this study can potentially help in evaluating the relation between the rainfall runoff model and satellite soil moisture. Exploiting this feedback is an essential step to help further develop the accuracy and applicability of such products for operational uses.

◯ SMOS SATELLITE

The SMOS mission is a joint program of the European Space Agency (ESA), the National Centre for Space Studies (CNES – Centre National d'Etudes Spatiales) and the Industrial Technological Development Centre (CDTI – Centro para el Desarrollo Technológico Industrial). The MIRAS instrument in the SMOS satellite is a dual polarized 2-D interferometer, which acquires data at the frequency of 1.4 GHz (L-band) designed to provide global information on surface soil moisture with an accuracy of 4% (Kerr et al. 2006; Kerr et al. 2001). In this study, Level 2 SMOS soil moisture products are used with detailed information given in the SMOS level 2 processor soil moisture algorithm theoretical basis document (ATBD) (Kerr et al. 2006).

The SMOS soil moisture products are defined on the ISEA 4H9 grid i.e. Icosahedral Snyder Equal Area projection (Pinori et al. 2008), defined on DGG (Discrete Global Grid). The radiometric resolution of the instrument is ~ 40 km with the soil moisture retrieval unit in $m^3 \, m^{-3}$ (i.e. volumetric). For building the model between the PDM SMD and SMOS soil moisture, the SMOS pixel with its centroid over the catchment is extracted using the Beam 4.9 package with SMOS 2.1.3 plugin and considered for the subsequent analysis.

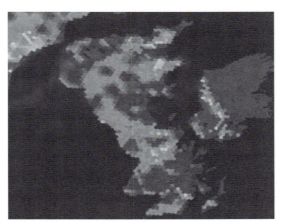

Figure 1 Sample SMOS soil moisture image over United Kingdom (Pin indicates the Brue catchment).

○ PROBABILITY DISTRIBUTED MODEL (PDM) ══════════

For a long time, rainfall–runoff models were used for flow prediction in catchments. These models are commonly used tools to extrapolate stream flow time-series in time and space for operational and scientific investigations (Ahmad et al. 2010). The lumped rainfall-runoff models require basically two types of data sources that are ETo/temperature and rainfall for flow prediction. Most of the model structures currently used for continuous modeling can be classified as conceptual based on two criteria (Ahmad et al. 2010) that are: (1) the structure of these models is specified prior to any modeling being undertaken and (2) the model parameters do not have a direct physical interpretation (being independently measurable) and estimated through calibration against observed data. Some of the well know conceptual models in hydrological modeling are IHACRES (Ahmad et al. 2010), PDM (Moore 2007), Stanford Watershed Model IV (CEH 2005), the SACRAMENTO model (Huang et al. 1996), XINANJIANG model (Clark and Arritt 1995) and some others.

As mentioned, there are many hydrological models available around the world and in this study a typical rainfall runoff model called PDM (Probability Distributed Model) is used. The PDM model is a fairly general conceptual rainfall-runoff model which transforms rainfall and evaporation data to flow at the catchment outlet and is well tested (Liu and Han 2010a; Moore 2007). The PDM has been widely applied throughout the world, both for operational and design purposes (Bell and Moore 1998). It has evolved as a toolkit of model functions that together constitute a lumped rainfall-runoff model capable of representing a variety of catchment-scale hydrological behaviors (Srivastava et al. 2014a). The model formulations are well suited for automatic parameter estimations. For real-time flow forecasting applications, the PDM model is complemented by updating methods based on error prediction and state-correction approaches (Bell and Moore 2000). PDM model require three main inputs – ETo, rainfall and river flow of the Brue catchment and the output products are flow and SMD (Soil Moisture Deficit). The SMD is derived to determine the effect of drying on the catchment on the actual evapo-transpiration (ET). The SMD routine is based on (Moore 2007):

$$\frac{E_i'}{E_i} = 1 - \left\{ \frac{(S_{max} - S(t))}{S_{max}} \right\}^{b_e} \tag{1}$$

where $\dfrac{E_i'}{E_i}$ is the ratio of actual ET to potential ET; and $(S_{max} - S(t))$ is Soil Moisture Deficit; b_e is an exponent in the actual evaporation function; S_{max} is the total available storage and $S(t)$ is storage at a particular time t.

TABLE 1 The model parameters available in PDM (Moore 2007).

Symbol	Model parameter	Unit
f_c	Rainfall factor	None
c_{min}	Minimum store capacity	mm
c_{max}	Maximum store capacity	
b	Exponent of pare to distribution controlling spatial variability of store capacity	None
b_e	Exponent in actual evaporation function	None
k_1	Time constants of cascade of two linear reservoirs	Hour
k_2		
k_b	Baseflow time constant	Hour mm^{m-1}
k_g	Groundwater recharge time constant	Hour mm^{bg-1}
S_t	Soil tension storage capacity	mm
b_g	Exponent of recharge function	None
q_c	constant flow representing returns/abstractions	m^{3}s^{-1}
τ_d	time delay	Hour

◯ GENERALIZED LINEAR MODEL (GLM) ━━━━━━━

These are the large class of statistical models for relating responses to linear combinations of predictor variables (Lindsey 1997). One variable is considered to be an explanatory variable (x_i) and the other is considered to be a dependent variable (y_i) (Johnson and Wichern 2002). The basic form of linear regression can be represented as:

$$y_i = x_i b + e_i \tag{2}$$

where x_i is a vector of k independent predictors, $i = 1, \ldots, n$; y_i is a dependent variable, b is a vector of unknown parameters and the e_i is stochastic disturbances.

The main characteristics of GLM model are that it has a stochastic component, a systematic component and a link between the random and systematic components (McCullagh and Nelder 1989). In Equation 3 or in case of linear regression model $x_i b$ is an identity function of the mean parameter; on the other hand in GLM, it is governed by some *link* function. In this study, several family and link functions are attempted to see the performance towards SMD prediction given in Table 2.

TABLE 2   Family and Link function used in this study.

Family	Link functions
Binomial	(link = "logit")
Gaussian	(link = "identity")
Gamma	(link = "inverse")
Poisson	(link = "log")

◯ CASE STUDY CATCHMENT

The Brue catchment, located in the south-west of England with an area of 135.5 Km²) is chosen as the study area. The river gauging point of the catchment is located at Lovington. The average altitude of this catchment is 105 m AMSL. The lowland wet grassland of the catchment forms part of the unique landscape of the Somerset Levels and Moors and the region is internationally and nationally designated for its conservation and landscape value (Wang et al. 2006). The mean river flow of this catchment is approximately 2.45 cumecs and elevation ranges between 35 metres to 190 metres above sea level. An average value of rainfall of 867 mm is recorded in the catchment. The layout of the Brue catchment is shown in Fig. 2.

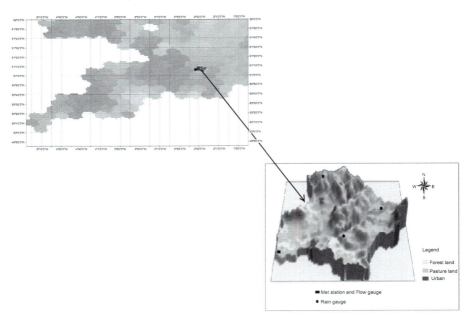

Figure 2   The layout of the Brue catchment.

○ PERFORMANCE STATISTICS

The performance of the PDM model is judged by the Nash–Sutcliffe efficiency (NSE) coefficients as an objective function (Nash and Sutcliffe 1970). It is based on the sum of the absolute squared differences between the simulated and observed values normalized by the variance of the observed values during the investigation period. The Nash-Sutcliffe Efficiency (NSE) is calculated using:

$$\text{NSE} = 1 - \frac{\sum_{i=1}^{n}[y_i - x_i]^2}{\sum_{i=1}^{n}[x_i - \bar{x}]^2} \qquad (3)$$

In this study we compared the various approaches that estimated SMD with the PDM SMD. The performances between two patterns are quantified in terms of the Coefficient of Determination (R^2), the Root Mean Square Error (RMSE) and the percentage of bias (%Bias). The percentage of bias measures the average tendency of the simulated values from the observed ones. The optimal value of %Bias is 0.0, with low-magnitude values indicating accurate model simulation (Eq. 4) with positive values indicating overestimation and vice versa. The R^2 is based on the sum of the absolute squared differences between the simulated and observed values normalized by the variance of the observed values during the investigation period. R^2 is calculated using Eq. 5 while Root Mean Square Error (RMSE) can be calculated using Eq. 6.

$$\%\text{Bias} = 100 * [\Sigma(y_i - x_i)/\Sigma(x_i)] \qquad (4)$$

$$R^2 = \left(\frac{n\Sigma xy - (\Sigma x)(\Sigma y)}{\sqrt{n(\Sigma x^2) - (\Sigma x^2)} \ \sqrt{n(\Sigma y^2) - (\Sigma y^2)}} \right)^2 \qquad (5)$$

$$\text{RMSE} = \sqrt{\left(\frac{1}{n}\sum_{i=1}^{n}[y_i - x_i]^2 \right)} \qquad (6)$$

where, x_i is observed, y_i is simulated flow and $\bar{x}$ is the mean.

○ RESULTS AND DISCUSSION

PDM Simulations for SMD Estimation

The calibration procedure of PDM model in particular focuses on the estimation of the optimal separation of slow (i.e. slow response system representing groundwater and other slow paths) and fast runoff (fast response system representing channel and other translational flow paths), and the good agreement between simulated and observed discharge. For the calibration of the PDM model, February 1, 2009 to January 31, 2011 i.e. two years of hourly datasets are used, while from February 1, 2011 to January 31, 2012, one year datasets are taken into account for validation. The performance of the model is judged by the Nash–Sutcliffe efficiency (NSE) coefficients as an objective function. In total, 13 parameters are taken into account for PDM initialization indicated in Table 2. The overall analysis indicates a satisfactory performance with the NSE value of 0.84 and 0.81 during the calibration and validation respectively. The hydrographs revealed that most of the time the model tends to match the hydrographs quite well with some low flow performance during April.

The main variable used in this study is soil moisture deficit which is used to distribute the rain to various types of runoff. In case of fast runoff, generally SMD is low and then a sizeable portion of runoff will occur as slow base flow groundwater runoff (Moore 2007). During the low flow time the model behavior is very unpredictable, can be visualized on the hydrograph which indicates that the model is unable to represent the nonlinear processes. However during most of the monitoring period the PDM model performance is relatively good. The time series between rainfall and flow during the calibration and validation period are shown in Fig. 3. The Sensitivity analysis (SA) and uncertainty analysis (UA) of the PDM model over the Brue catchment are well presented in detail by Srivastava et al. in (2014a).

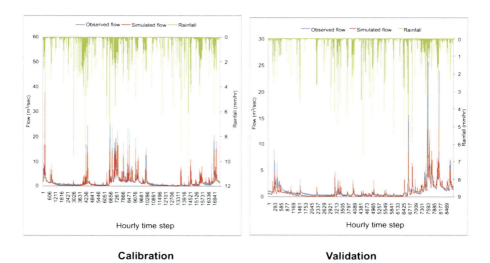

Calibration   **Validation**

Figure 3 Observed and Simulated flow during calibration and validation from PDM model with rainfall on secondary axis.

SMD and Soil Moisture Temporal Variation

The comparisons between soil moisture and PDM SMD time series exhibit a high temporal variability during January and February and follow a strong temporal cycle. A very high value of soil moisture is recorded during the period of December and January. As compared to soil moisture, the increasing temperatures and high evaporation through the period April-May to August-September lead to a progressive drying of the soil that leads to an increase in SMD values. When SMD rates slacken in November-December and rainfall wets-up the soil profile–a surging graph can be seen in Fig. 4. As expected, the soil moisture increases after a rain event and drying out follows an exponential decay. The analysis of the rainfall time series shows that March to May are slightly drier than other months while November-December are the wettest periods during the monitoring period. During the analysis, some moderately short-duration storms are observed which leads to maximum rainfall intensity during the month of June. Soil moisture is found lower over the winter months and near to field capacity until the mid of April during the most of the year. However, increasing temperature caused a substantial SMD development prominent from April to the beginning of August (usually, the driest and warmest period of the year). Generally, soil moisture is highly responsive with significant fluctuations over the entire period to the even smaller changes in SMD.

The scatter plots between SMOS soil moisture and PDM SMD datasets are depicted through Fig. 5 along with the estimated Pearson and Spearman correlations. The two correlation statistics are calculated in order to assure the linear and nonlinear complexity in the datasets. The Spearman ($r_{spearman}$) and Pearson ($r_{pearson}$) correlation statistics between SMOS soil moisture and PDM SMD yield nearly similar values which indicate that no strong nonlinearity existed between the datasets and a simple GLM could be useful for SMD prediction because of its less complexity (Blumer et al. 1987). As expected, a negative correlation obtained between the PDM SMD and SMOS datasets indicates an inverse relation between the products. A moderate correlation statistics is obtained between the datasets with the values of $r = -0.63$ and $rs = -0.61$.

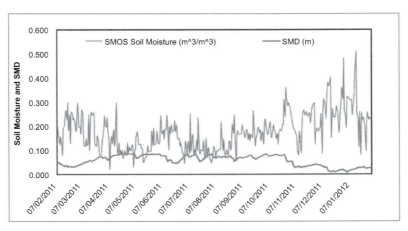

Figure 4 Temporal trend in SMOS soil moisture and PDM SMD.

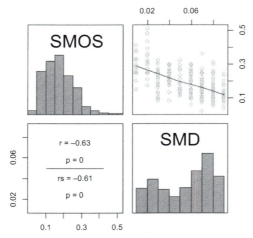

Figure 5 Spearman's and Pearson's correlation matrix plot between SMOS and PDM SMD.

Performances of GLM for SMD Prediction

The time series generated from all the approaches during the validation is indicated through Fig. 6. The comparisons between the simulated time series and PDM SMD exhibit a close relationship with high temporal variability during the monitoring period. During the period, the trend and pattern in the datasets are very close to the one simulated by the PDM model. The distinction between volumetric water content and the degree of soil saturation is often unclear in the literature (Miller et al. 2007), and both terms are used here to describe the variations in SMD. The time series charted the course of the daily SMD simulated from GLM over 4 months of the datasets. From these, daily variability and seasonal patterns SMD could be determined. The distinctive dry periods can be demonstrated, when a prominent rise in the SMD occurred in the figure. This pattern indicates that because of SMD, the catchment don't have appreciable amount of water in the soil profile and fluctuation may occur quickly after a change in rainfall amount.

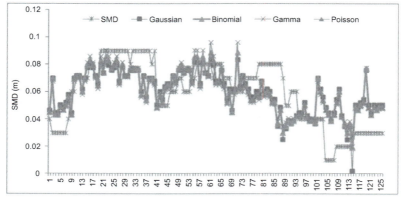

Figure 6 Time series depicting the pattern between GLM simulations and PDM SMD.

For developing an empirical relationship, the GLM model using the PDM SMD and SMOS soil moisture are used with different family and link functions as shown in Table 2 while the performance statistics obtained from different techniques are indicated through Table 3. For deriving the empirical relationship, the calibration and validation datasets are first divided into two lots. From each month, two thirds of the data are taken as calibration and the other third as validation, so that both calibration and validation data are representative of all the seasons. After estimating the empirical relationships, the validation results are used to test the algorithms. The scatter plots obtained from all the approaches are depicted through Figure 7 along with the Square of Correlation (R^2) values. The square of correlation during the validation indicated a value of $R^2 = 0.421$ in case of Binomial and Poisson followed by the Gaussian ($R^2 = 0.409$) and Gamma ($R^2 = 0.402$), which indicates that the Binomial and Poisson are promising for simulation of SMD from SMOS soil moisture with marginally high performance as compared to the other techniques.

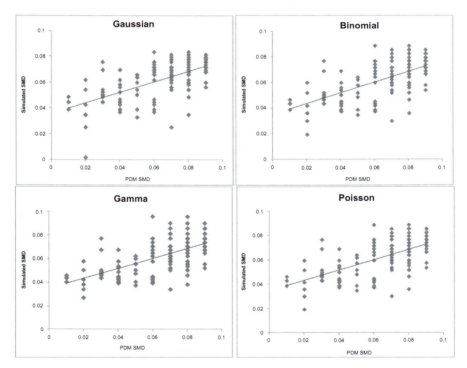

Figure 7 Scatterplots representing the performances of different approaches during the validation.

For a diagnostic performance evaluation of the model, two performance statistics- RMSE and %Bias are estimated to compare the PDM SMD with the GLM simulated datasets. Looking at the values of the %Bias, the Gamma with inverse link function performed well with least overestimation, although the R^2 and RMSE are lesser than the other techniques. Highest %Bias is indicated by Poisson with log link function with least RMSE and R^2 in comparison to the other techniques such as Gaussian, Binomial and Gamma. The analysis indicates that the outputs are comparable, although some higher performance are reported by the techniques Binomial and Poisson in terms of R^2 and RMSE. However, they are not good in reducing the %Bias of the datasets, which end up with slight overestimation. The overall performance of GLM with different family and link functions showed that the Binomial with identity, and Poisson with log link function can be used for estimation of SMD from SMOS soil moisture.

TABLE 3 R^2, %Bias and RMSE Statistics during validation.

Family/link functions	R^2	% Bias	RMSE
Gaussian/logit	0.409	0.9065	0.0182
Binomial/identity	0.421	0.9304	0.0180
Gamma/inverse	0.402	0.8800	0.0183
Poisson/log	0.420	0.9320	0.0180

CONCLUSION

Soil moisture has proven its importance for hydrological sciences and considered as a key variable in the water and energy exchanges that occur at the land-surface/atmosphere interface. On the other hand, SMD is an integral component for most of the rainfall-runoff models and needed for designing the water balance equations. Satellite soil moisture such as from SMOS is promising for SMD estimation using empirical relationship estimated through GLM model using different linear regression families and link functions. The overall performance of GLM in terms of coefficient of determination, RMSE and %Bias showed that the Binomial with identity and Poisson with log link function can be used for estimating SMD from SMOS soil moisture. However, more research in the technical literature domain on SMD is required, so that useful knowledge and experiences can be accumulated over the different geographical areas and by using the other promising satellites such as SMAP, Aquarius, or AMSR-E/2.

ACKNOWLEDGMENT

The authors would like to thank the Commonwealth Scholarship Commission, British Council, United Kingdom and Ministry of Human Resource Development, Government of India for providing the necessary support and funding for this research. The authors are highly thankful to the European Space Agency for providing the SMOS data. The authors would like to acknowledge the British Atmospheric Data Centre, United Kingdom for providing the ground observation datasets. The first author would like to thanks the SMOS team for providing the training regarding

SMOS at CESBIO, Toulouse, France. The authors also acknowledge the Advanced Computing Research Centre at the University of Bristol for providing the access to the supercomputer facility (The Blue Crystal).

○ REFERENCES

Ahmad, S., Kalra, A. and Stephen, H. 2010. Estimating soil moisture using remote sensing data: a machine learning approach. Advances in Water Resources 33: 69-80.

Al Bitar, A., Leroux, D., Kerr, Y.H., Merlin, O., Richaume, P., Sahoo, A. and Wood, E.F. 2012. Evaluation of SMOS soil moisture products over continental US using the SCAN/SNOTEL network. IEEE Transactions on Geoscience and Remote Sensing, 50: 1572-1586.

Bell, V. and Moore, R. 1998. A grid-based distributed flood forecasting model for use with weather radar data: Part 2. Case studies Hydrology and Earth System Sciences 2: 283-298.

Bell, V.A. and Moore, R.J. 2000. The sensitivity of catchment runoff models to rainfall data at different spatial scales. Hydrology and Earth System Sciences Discussions 4: 653-667.

Beven, K. and Wood, E.F. 1983. Catchment geomorphology and the dynamics of runoff contributing areas. Journal of Hydrology 65: 139-158.

Blumer, A., Ehrenfeucht, A., Haussler, D. and Warmuth, M.K. 1987. Occam's razor. Information processing letters 24: 377-380.

Cabus, P. 2008. River flow prediction through rainfall–runoff modelling with a probability-distributed model (PDM) in Flanders, Belgium. Agricultural Water Management 95: 859-868.

Calder, I., Harding, R. and Rosier, P. 1983. An objective assessment of soil-moisture deficit models. Journal of Hydrology 60: 329-355.

CEH, 2005. A Guide to the PDM Rainfall Runoff Model. Centre for Ecology & Hydrology, Wallingford, 22 edition.

Clark, C.A. and Arritt, P.W. 1995. Numerical simulations of the effect of soil moisture and vegetation cover on the development of deep convection. Journal of Applied Meteorology 34: 2029-2045.

Croke, B.F. and Jakeman, A.J. 2004. A catchment moisture deficit module for the IHACRES rainfall-runoff model. Environmental Modelling & Software 19: 1-5.

Evans, J. and Jakeman, A. 1998. Development of a simple, catchment-scale, rainfall-evapotranspiration-runoff model. Environmental Modelling & Software 13: 385-393.

Huang, J., van den Dool, H.M. and Georgarakos, K.P. 1996. Analysis of model-calculated soil moisture over the United States (1931-1993) and applications to long-range temperature forecasts. Journal of Climate 9: 1350-1362.

Jackson, T.J. 1993. III. Measuring surface soil moisture using passive microwave remote sensing. Hydrological Processes 7: 139-152.

Jackson, T.J., Le Vine, D.M., Swift, C.T., Schmugge, T.J. and Schiebe, F.R. 1995. Large area mapping of soil moisture using the ESTAR passive microwave radiometer in Washita'92. Remote Sensing of Environment 54: 27-37.

Jackson, T.J., Le Vine, D.M., Hsu, A.Y., Oldak, A., Starks, P.J., Swift, C.T., Isham, J.D. and Haken, M. 1999. Soil moisture mapping at regional scales using microwave radiometry: The Southern Great Plains Hydrology Experiment. IEEE Transactions on Geoscience and Remote Sensing 37: 2136-2151.

Jackson, T.J., Bindlish, R., Cosh, M.H., Tianjie, Z., Starks, P.J., Bosch, D.D., Seyfried, M., Moran, M.S., Goodrich, D.C., Kerr, Y.H., and Leroux, D. 2012. Validation of Soil Moisture and Ocean Salinity (SMOS) Soil Moisture Over Watershed Networks in the U.S. IEEE Transactions on Geoscience and Remote Sensing 50: 1530-1543.

Johnson, R.A. and Wichern, D.W. 2002. Applied Multivariate Statistical Analysis, Vol. 4. Prentice Hall, Upper Saddle River, NJ.

Kerr, Y.H., Waldteufel, P., Wigneron, J.P., Martinuzzi, J., Font, J. and Berger, M. 2001. Soil moisture retrieval from space: The Soil Moisture and Ocean Salinity (SMOS) mission. IEEE Transactions on Geoscience and Remote Sensing 39: 1729-1735.

Kerr, Y., Waldteufel, P., Richaume, P., Davenport, I., Ferrazzoli, P. and Wigneron, J. 2006. SMOS level 2 processor soil moisture algorithm theoretical basis document (ATBD) SM-ESL (CBSA), CESBIO, Toulouse, SO-TN-ESL-SM-GS-0001, V5 a, 15/03.

Lindsey, J.K. 1997. Applying Generalized Linear Models. Springer Verlag, New York.

Liu, J. and Han, D. 2010a. Indices for calibration data selection of the rainfall-runoff model. Water Resources Research 46: W04512.

Liu, J. and Han, D. 2010b. Indices for calibration data selection of the rainfall-runoff model. Water Resources Research. DOI: 10.1029/2009WR008668.

McCullagh, P. and Nelder, J.A. 1989. Generalized Linear Models. Chapman & Hall/CRC, Florida.

Miller, G.R., Baldocchi, D.D., Law, B.E. and Meyers, T. 2007. An analysis of soil moisture dynamics using multi-year data from a network of micrometeorological observation sites. Advances in Water Resources 30: 1065-1081.

Mladenova, I., Lakshmi, V., Jackson, T.J., Walker, J.P., Merlin, O. and de Jeu, R.A. 2011. Validation of AMSR-E soil moisture using L-band airborne radiometer data from National Airborne Field Experiment 2006. Remote Sensing of Environment 115(8): 2096-2103.

Moore, R.J. 2007. The PDM rainfall-runoff model. Hydrology and Earth System Sciences Discussions 11: 483-499.

Moran, M., Clarke, T., Inoue, Y. and Vidal, A. 1994. Estimating crop water deficit using the relation between surface-air temperature and spectral vegetation index. Remote Sensing of Environment 49: 246-263.

Narasimhan, B. and Srinivasan, R. 2005. Development and evaluation of Soil Moisture Deficit Index (SMDI) and Evapotranspiration Deficit Index (ETDI) for agricultural drought monitoring. Agricultural and Forest Meteorology 133: 69-88.

Narsimlu, B., Gosain, A.K., Chahar, B.R., Singh, S.K. and Srivastava, P.K. 2015. SWAT model calibration and uncertainty analysis for streamflow prediction in the Kunwari River Basin, India, using sequential uncertainty fitting. Environmental Processes 2: 79-95.

Nash, J.E. and Sutcliffe, J. 1970. River flow forecasting through conceptual models part I—A discussion of principles. Journal of Hydrology 10: 282-290.

Patel, D.P. and Srivastava, P.K. 2013. Flood hazards mitigation analysis using remote sensing and GIS: correspondence with town planning scheme. Water Resources Management 27(7): 2353-2368.

Patel, D.P. and Srivastava, P.K. 2014. Application of geo-spatial technique for flood inundation mapping of low lying areas. pp. 113-130. *In*: Srivastava, P.K., Mukherjee, S., Gupta, M. and Islam, T. (eds.). Remote Sensing Applications in Environmental Research. Springer, Switzerland.

Petropoulos, G.P., Ireland, G., Srivastava, P.K. and Ioannou-Katidis, P. 2014. An appraisal of the accuracy of operational soil moisture estimates from SMOS MIRAS using validated in situ observations acquired in a Mediterranean environment. International Journal of Remote Sensing 35: 5239-5250.

Pinori, S., Crapolicchio, R. and Mecklenburg, S. 2008. Preparing the ESA-SMOS (soil moisture and ocean salinity) mission-overview of the user data products and data distribution strategy. IEEE, pp 1-4.

Rushton, K., Eilers, V. and Carter, R. 2006. Improved soil moisture balance methodology for recharge estimation. Journal of Hydrology 318: 379-399.

Schlenz, F., Loew, A. and Mauser, W. 2012. First results of SMOS soil moisture validation in the Upper Danube Catchment. IEEE Transactions on Geoscience and Remote Sensing, 50: 1507-1516.

Smith, J.M. 1986. Mathematical Modelling and Digital Simulation for Engineers and Scientists. John Wiley & Sons, Inc., Michigan.

Srivastava, P.K., Han, D. and Rico-Ramirez, M.A. 2012. Assessment of SMOS satellite derived soil moisture for soil moisture deficit estimation. *In:* Prediction in Ungauged basin (PUBS) symposium, Delft University of Technology, Delft, Netherlands and IAHS, 22-25 October 2012.

Srivastava, P.K. 2013. Soil Moisture Estimation from SMOS Satellite and Mesoscale Model for Hydrological Applications. PhD Thesis, University of Bristol, Bristol, United Kingdom.

Srivastava, P.K., Han, D., Ramirez, M.R. and Islam, T. 2013a. Appraisal of SMOS soil moisture at a catchment scale in a temperate maritime climate. Journal of Hydrology 498: 292-304.

Srivastava, P.K., Han, D., Ramirez, M.R. and Islam, T. 2013b. Machine learning techniques for downscaling SMOS satellite soil moisture using MODIS land surface temperature for hydrological application. Water Resources Management 27: 3127-3144.

Srivastava, P.K., Han, D., Rico-Ramirez, M.A., Al-Shrafany, D. and Islam, T. 2013c. Data fusion techniques for improving soil moisture deficit using SMOS satellite and WRF-NOAH land surface model. Water Resources Management 27: 5069-5087.

Srivastava, P.K., Han, D., Rico-Ramirez, M.A. and Islam, T. 2014a. Sensitivity and uncertainty analysis of mesoscale model downscaled hydro-meteorological variables for discharge prediction. Hydrological Processes 28: 4419-4432.

Srivastava, P.K., Han, D., Rico-Ramirez, M.A., O'Neill, P., Islam, T. and Gupta, M. 2014b. Assessment of SMOS soil moisture retrieval parameters using tau–omega algorithms for soil moisture deficit estimation. Journal of Hydrology 519: 574-587.

Tripp, D.R. and Niemann, J.D. 2008. Evaluating the parameter identifiability and structural validity of a probability-distributed model for soil moisture. Journal of Hydrology 353: 93-108.

Zhuo, L., Han, D., Dai, Q., Islam, T. and Srivastava, P.K. 2015. Appraisal of NLDAS-2 Multi-Model simulated soil moistures for hydrological modelling. Water Resources Management 29(10): 3503-3517.

Section IV

Artificial Intelligence and Hybrid Approaches

A Deterministic Model to Predict Frost Hazard in Agricultural Land Utilizing Remotely Sensed Imagery and GIS

Panagiota Louka,[1,]* *Ioannis Papanikolaou,*[1] *George P. Petropoulos*[2] and *Nikolaos Stathopoulos*[3]

ABSTRACT

Frost risk is a critical factor in agricultural planning and management and has a significant impact on the environment and society. The ability to map the spatio-temporal distribution of frost conditions over a given area has improved dramatically nowadays with the rapid technological developments in Earth Observation (EO) technology and Geographical Information Systems (GIS).

In this study, a deterministic frost risk mapping model for agricultural crops cultivated in Mediterranean environments is proposed. The model is based on the main factors that govern frost risk including environmental parameters such as land surface temperature and geomorphology. Its implementation is based primarily on Earth Observation (EO) data from MODIS and ASTER polar orbiting sensors, supported also by ancillary ground observation data. Topographical parameters required in the model include the altitude, slope, steepness, aspect, topographic curvature and extent of the area influenced by water bodies. Additional data required include land use and vegetation classification (i.e. type and density).

[1] Department of Natural Resources Development and Agricultural Engineering, Agricultural University of Athens, Greece.
[2] Department of Geography & Earth Sciences, University of Aberystwyth, Wales, UK.
[3] Section of Geological Sciences, School of Mining Engineering, National Technical University of Athens, Greece.
* Corresponding author: p.louka@aua.gr

The model has been tested in a region of primarily agricultural land located towards the northwestern part of the Greek mainland. The model's ability to predict frost conditions for the winter period was validated for four selected years (i.e. 2004, 2006, 2009 and 2010). In addition, model predictions were compared against ground observations of frost damage in agricultural crops acquired from field surveys conducted by the Greek Agricultural Insurance Organization.

The results obtained verified the ability of the model to produce reasonable estimations of the spatio-temporal distribution of frost conditions in our study area, following the largely explainable patterns in respect to the study site's geomorphological and local weather condition characteristics. Overall, the proposed methodology proved to be capable of detecting frost risk in Mediterranean environments in a time efficient and cost effective way, making it a potentially very useful tool for agricultural management and planning. The proposed model may be used as an important tool for frost mapping, a natural hazard that leads to severe vegetation damage and agricultural losses.

KEYWORDS: Remote sensing, GIS, frost, MODIS, risk assessment, Greece.

○ INTRODUCTION

Frost has been globally identified as a leading agricultural hazard, as it can occur in almost any location, outside the tropical zones (FAO 2005). About two thirds of the world's landmass is annually subjected to temperatures below the freezing point and about half of it suffers from temperatures below $-20°C$ (Larcher 2001). Frost can be described as a meteorological event when crops and other plants experience freezing injury due to occurrence of an air temperature less than $0°C$ (FAO 2005).

Huge economic losses caused by freezing of sensitive horticultural crops are observed throughout the world. For example, the citrus industry in Florida has been occasionally devastated by frost damage (Cooper et al. 1964; Martsolf et al. 1984; Attaway 1997), coffee production in Brazil has been seriously damaged by frost (Hewitt 1983) and winterkill of cereals is a major problem (Stebelsky 1983; Caprio & Snyder 1984a; Cox et al. 1986). Apart from the economic losses, there are also significant secondary effects on local and regional communities, such as unemployment (FAO 2005).

Many countries in temperate and arid climates experience economic and social problems due to frost damage, especially at high elevations (FAO 2005). Even in a Mediterranean country, such as Greece, compensations for frost damage in crops reach to an amount of 82 million euro per year, which is 32% of the total agricultural compensations (ELGA 2000-2010 data). Frost damage can be a serious hazard for many types of temperate climate crops. It may damage leaves and fruit and have significant impact on plants' health (even destroy it). The level of frost injuries on a plant depends on the severity of the frost event and on the particular plant's susceptibility (Richards 2003).

Minimum land temperature distribution in an area controls vegetation zonation and productivity of plants. It also affects most biotic processes such as plant phenology, growth, evapotranspiration, moisture requirements, carbon fixation and decomposition in natural and cultivated mountain ecosystems (Kramer 1983; Waring & Schlesinger 1985; Aber & Melillo 1991; Chen et al. 1999). Tree growth, species composition and plants' susceptibility to disturbance are impacted cumulatively in a significant level (Turner & Gardner 1991).

Depending on vegetation structure, landscape position or soil properties, frost can damage plant tissues thus affecting forest, pasture and crop productivity (Blennow & Lindkvist 2000). Frost may cause scorching and browning of young leafs and buds, fruit and flower drop, development of misshapen fruits, as presented in Figures 1(a)-(f) (ELGA – GAIO 2003).

Figure 1 Frost damage on apple trees in the stage of (a) swollen buds, (b) bud burst, (c) green cluster, (d) bloom, (e) petal fall and (f) fruit set (ELGA 2003).

In the same plant, resistance to frost varies amongst its tissues. Meristematic cells are in general less frost resistant than mature tissues (Sakai & Larcher 1987). The extent of frost damages on plant tissues depend on factors such as the plant's developmental stage, as well as the duration and severity of frost and the rates of cooling. Because of cellular water freezing, frost imposes stress on plants due to the dehydration of plant tissues and cells. The effects of frost in plants depend on whether ice formation takes place intracellularly or extracellularly in the intercellular spaces (Beck et al. 2004).

Spatio-temporal Distribution of Frost Events

Through the recent decades, considerable effort has been expended in mapping the spatiotemporal resolution of minimum surface temperatures, frost events and the estimation of frost risk.

Being able to acquire information and map the extent of past frost events has been underlined as a key aspect to both environmental scientists and policy makers. Accurate and rapid mapping of the geographical and temporal distribution of the frosted area and of the damage occurred from frost conditions allows estimating the economic consequences from such a natural hazard and establishing rapid rehabilitation and restoration policies in the damaged areas. It is a very significant tool in successful agricultural management, decision making and regional management. It is important to have information concerning not only the temporal, but also the spatial distribution of frost hazard (e.g. Chuanyan et al. 2005).

Nowadays, the rapid technological and scientific development particularly in Earth Observation (EO) and Geographic Information Systems (GIS) has provided a great potential in the study of frost events. Satellite imagery provides information on the spatiotemporal distribution of surface temperature, covering large areas, parts of which may be inaccessible by other means. These data also provide the ability to study the effect of physiographic factors in frost occurrence.

Various studies have manifested the importance of several factors as being of critical importance in accurately estimating frost hazard on a regional scale, some of the most important ones being the following:

Elevation: Elevation is a key factor in surface temperature variance, depending on atmospheric stability. Surface temperature declines by 0.65°C per 100 m, with the exception of temperature inversions (Shao et al. 1997; Pouteau et al. 2011). It has been observed that as altitude rises, mean minimum temperature tends to be lower; therefore the mean annual frost free days tend to be fewer (Geiger et al. 2003).

Slope Gradient: The topography of an area and especially the slope angle and orientation are considered important factors that determine surface temperature. Sloping areas display a lower frost hazard due to cold air drainage towards areas of lower elevation. Relatively flat areas, especially in low-lying lands, are more susceptible to frost than slopes situated even in altitudes of 4,000 m (Radcliffe & Lefever 1981; Fridley 2009; Pouteau et al. 2011).

Slope Aspect: The orientation of slopes has a significant influence in the climatic conditions of an area. In slopes facing west to north, the incoming potential insolation is reduced and there are lower temperatures and greater intensity and duration of snowfall, leading to greater frost risk levels (Radcliffe & Lefever 1981; Fridley 2009; Pouteau et al. 2011).

Topographic Curvature: Due to cold air drainage, lower parts of valleys are often more susceptible to frost events as cold air masses tend to accumulate causing extensive crop damages. Concave areas are generally more susceptible to frost than convex (Soderstrom et al. 1995; ELGA 2003).

Hydrographic Network: The presence of water bodies plays a role in protection from frost hazard. Especially large water bodies such as lakes, major rivers and the sea, significantly influence the temperature of cold air masses and reduce the intensity of advection frosts (ELGA 2003; Fridley 2009; Pouteau et al. 2011). For example, Pouteau et al. (2011) found that the factors with the most prominent effects are altitude and distance to salt lakes at large scales, whereas slope, topographic convergence and insolation gained influence at local scales (Pouteau et al., 2011).

Compound Topographic Index: The Compound Topographic Index[*] (CTI) comprises one of the most important indexes of cold air drainage. Low CTI values represent convex conditions and low frost hazard levels whereas high CTI values represent concave conditions like coves or hill slope bases, where cold air has the tendency to accumulate and frost hazard is higher (Gessler et al. 2000; Pouteau et al. 2011).

Land Use and Cover: Land use and land cover is proven to have significant role in the climate of an area. In general, results of various studies suggested that in urban areas, water bodies, or densely forested areas fewer frost incidents occur, due to higher temperatures (compared to surrounding areas) in winter time (Gustavsson et al. 1998; Chapman & Thornes 2006).

Temperature: To predict the temperature variation at the watershed or regional scale, methods such as the vertical lapse method, regional regression and geostatistical techniques have been developed (Leffler 1981; Boyer 1984; Russo et al. 1993; Régnière & Bolstad 1994; Régnière 1996; Bolstad et al. 1997). Until fairly recently, long term mean temperature data were recorded and used to produce topoclimate maps, contoured 'by hand' using expert knowledge (Richards 2003). Yet, a shortfall of these methods is that they do not incorporate the terrain effect in temperature variation, which is based on the different annual amount of sun radiation falling on a unit area of the surface (Bolstad et al. 1998).

The implementation of GIS with EO data also enhances the ability to study the effect of topographic factors, such as slope angle and orientation, in temperature variation at fine spatial scales over arbitrary periods of time (Dubayah 1994; Dubayah & Rich 1995; Safanda 1999). Indeed, several researchers have computed the correlation between satellite derived surface temperatures and measurements of air temperature with suitable long-term data. The synergistic use of satellite-derived data with meteorological data has also led to allowing the extrapolation of important historical records over large horizontal areas (Kerdiles et al. 1996; Francois et al. 1999; Domenikiotis et al. 2003).

The current study provides further evidence of the advantages of combining data from independent sources such as satellite data, GIS and ground observations for the mapping of temperature variation. In particular the objectives of the present work are (i) to build a frost risk prediction model based on local terrain features such as aspect and slope gradient, vegetation density and land use variation; (ii) to evaluate the method by developing a frost frequency map using satellite MODIS Land Surface Temperature data; and (iii) to provide a method for the identification of high frost risk zones in the study area utilizing ground observations of frost damages in agricultural areas to further validate the results of the frost risk model.

---

[*] CTI is a steady state wetness index. The CTI is a function of both the slope and the upstream contributing area per unit width orthogonal to the flow direction. The implementation of CTI can be shown as: CTI = ln (As/(tan(beta))), where As = Area Value calculated as (flow accumulation + 1) * (pixel area m^2) and beta is the slope expressed in radians (Evans J. 2004).

○ EXPERIMENTAL SET UP

Study Area

The study area is situated in northwestern Greece and is comprised of almost 500 municipal districts in 11 different prefectures. The landscape of the area can be described as a mosaic of extended flat shores in seaside regions and an alternation of valleys and mountainous areas culminating at 2,267 m. There are three types of climatic conditions: Mediterranean in the maritime zones, warm temperate in continental parts and sub-arctic with cool summer in the mountains.

The hydrographic network has a high density throughout the study area due to impermeable geological formations of flysch and Neogene marls which increase surface runoff. There in an extensive drainage network involving 3 rivers and 2 lakes. In the northeastern part, especially, there is a significantly increased drainage density, as 4 rivers flow into the sea (Fig. 2).

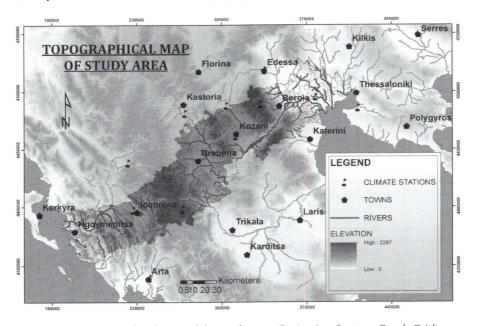

Figure 2 Topographical map of the study area (Projection System: Greek Grid).

In order to study the alterations of the climatic conditions of the research area, 30-year data sets of 11 climate stations installed within or near the area of interest were processed (Table 1).

The values of important climatic parameters (such as temperature, rainfall, etc.) of the study area were interpolated using the Inverse Distance Weighted Algorithm in GIS (Figures 3a-d). Data was acquired from the Hellenic National Meteorological Service and for the years 1955 to 1997.

TABLE 1 Climate stations installed within or near the study area, by the Hellenic National Meteorological Service and the coordinates of their position in Geodetic Reference System 1980 (GWS80).

Climate station name	Latitude	Longitude	Altitude (m)
Arta-city	39°10′	20°59′	39.0
Arta-filothei	39°10′	21°00′	11.5
Edessa	40°48′	22°03′	237.0
Ioannina	39°40′	20°51′	483.0
Kastoria	40°27′	21°17′	660.0
Kozani	40°17′	21°47′	625.0
Konitsa	40°03′	20°45′	542.0
Mikra	40°31′	22°58′	4.0
Ptolemaida	40°31′	21°41′	601.0
Sedes	40°32′	22°01′	51.9
Trikala Imathias	40°36′	22°33′	5.8

(a)

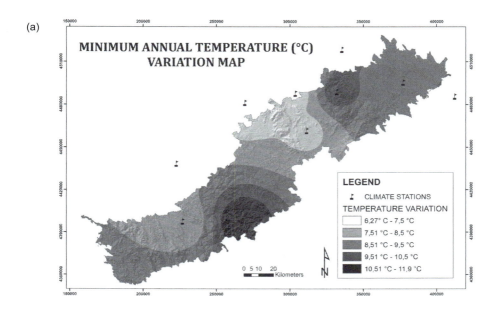

(b)

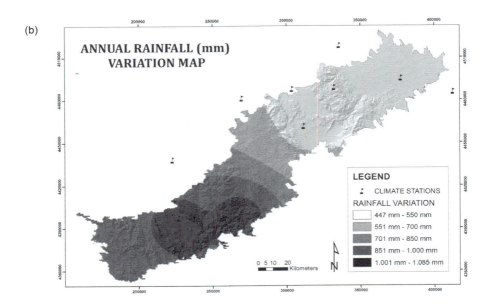

(c)

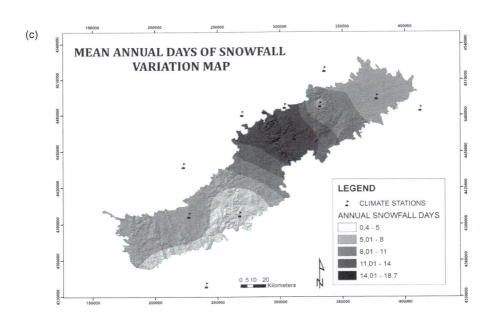

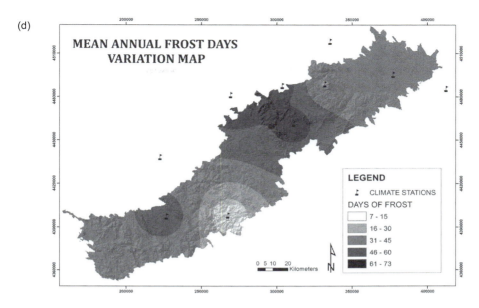

Figure 3 Variation of climatic features in the study region, based on meteorological data of 11 available climate stations (years 1955-1997) (a) annual mean minimum temperature, (b) annual rainfall, (c) mean annual snow days, (d) mean annual frost days (Projection System: Greek Grid).

According to the acquired meteorological data, mean minimum annual temperature varies from 5.9°C to 11.9°C. The coldest regions of the study area, according to the map presented in Fig. 3a, are observed in the northwestern and southwestern areas near the central part. Mean annual precipitation varies between 448 mm and 1,081 mm. Fig. 3b show the rainfall distribution along the study area, illustrates that higher values of rainfall are observed in the southeastern part of the study area.

The interpolation results of the climatic data suggest that snowfall is more intense in the central and more elevated parts of the study area, as exhibited in Fig. 3c. Mean annual days of snowfall in the study area vary between 0.3 and 19.6 days, whereas mean annual days of frost vary between 7 and 73 days. The most frost affected areas according to these 11 stations are situated in the northern and eastern part of the central study area, as presented in Fig. 3d.

Datasets

For the current study, different types of data were selected and processed. Table 2 provides a summary of all the relevant datasets used in this analysis including the data source as well as the details of the derived information. All data used in this study were available at no cost.

TABLE 2 Data sets of present study.

Data	Source	Information derived
ASTER GDEM, V. 2	https://lpdaac.usgs.gov	Elevation Slope gradient and aspect Slope curvature
Hydrological network	www.geodata.gov.gr	Distance from water bodies
CORINE 2000, V. 1	www.geodata.gov.gr	Land use & cover
MODIS LST	https://lpdaac.usgs.gov	Surface temperature
Ground observations	Agricultural Insurance Organization of Greece (ELGA)	Frost damage in farms
Meteorological data	Hellenic National Meteorological Service (EMY)	Temperature, rainfall, days of snow and frost

A Digital Elevation Model (DEM) with a 30 m spatial resolution and mean accuracy was downloaded at no cost from the ASTER GDEM website. The mean accuracy of the ASTER GDEM has been found to be 20 m (Gesch et al. 2012) and according to two previous studies concerning areas in Greece it has been estimated to be 12.41 and 13.6 m (Chrysoulakis et al. 2004; Miliaresis & Paraschou 2011).

Additionally, MODIS land surface temperature images (MOD11A1) with 1 km resolution were downloaded from the MODIS web site. They are level L3 products in tiles of 1.113 km^2 and have accuracy of 1°K (in land and under clear sky conditions). In order to cover the whole area of interest two MODIS tiles were needed (column 19 and row 4 and 5). In total 1,280 images were downloaded which correspond to 640 different dates of the years 2004, 2006, 2009 and 2010.

Meteorological data including mean minimum annual temperature, mean annual rainfall, mean annual days of snowfall and frost were obtained by the Hellenic National Meteorological Service. These data cover a period of 30 years, specifically years 1955-1997 (HNMS 1999).

Finally, the processing of ground observation data, concerning frost damage on agricultural yields during the years 1999-2010, was accomplished.

○ METHODOLOGY

The methodology which was adopted in the present study consisted of three basic stages outlined below (and also summarized in Fig. 4):

Step 1: Development of a frost hazard model in a GIS environment.

Step 2: Development of a frost frequency map product based on multi-temporal analysis of MODIS land surface temperature data.

Step 3: Mapping of frost damage distribution on agricultural land, based on ground observations by Agricultural Insurance Organization of Greece (ELGA).

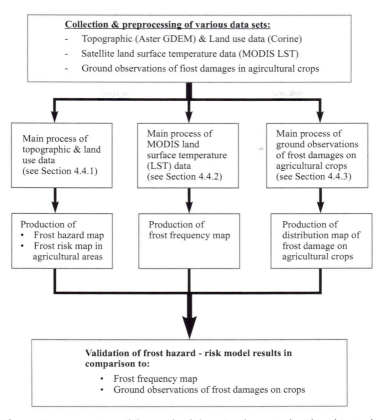

Figure 4 An overview of the methodology implemented within this study.

Development of Frost Hazard Model

In order to map frost hazard in the study area, the methodology of weighted linear combination and ordered weighted averaging has been adopted. This is regarded as one of the most important methods for producing complex hazard maps in a GIS environment (Drobne & Lisec 2009; Malczewski 2011).

First of all, a rationale was devised which sets out how topography, vegetation and land use influence temperature fluctuation and incidents of frost. On the basis of the relevant of literature review (Section 1.1), the key model factors were selected in developing the frost hazard model of this study.

Secondly, suitable datasets were identified and collected from different sources (see Table 1). The background assumption of which factors/datasets are correlated with frost events is based on scientific literature. After the pre-processing implementation, thematic maps with a spatial resolution of 100 m of the selected frost hazard factors were created.

The thematic layers were classified into categories and were assigned a weight factor according to studies that evaluate the effect of each parameter on frost events (Radcliffe & Lefever 1981; Soderstrom et al. 1995; Gustavsson et al. 1998; Gessler et al. 2000; Geiger et al. 2003; Chapman & Thornes 2006; Fridley 2009; Pouteau et al. 2011).

TABLE 3 Classification of frost hazard factors.

Factor	Weight	Ranks	Classes
Elevation	30 %	5	1,300 m – 2,267 m
		4	900 m – 1,300 m
		3	600 m – 900 m
		2	250 m – 600 m
		1	0 m – 200 m
Slope aspect	10 %	5	337.5 – 360.0 & 0.0 – 22.5
		4	22.5 – 67.5 & 292.5 – 337.5
		3	67.5 – 112.5 & 247.5 – 292.5
		2	112.5 – 157.5 & 202.5 – 247.5
		1	157.5 – 202.5
Slope gradient (°)	10 %	5	0° – 5°
		4	5° – 11°
		3	11° – 18°
		2	18° – 26°
		1	< 26°
Curvature	10 %	5	Curvature from –2.77 to –0.5
		4	Curvature from –0.5 to –0.1
		3	Curvature from –0.1 to 0.1
		2	Curvature from 0.1 to 0.5
		1	Curvature from 0.5 to 2.75
C.T.I.	15 %	5	14.0 – 24.92
		4	11.0 – 14.0
		3	8.0 – 11.0
		2	6.0 – 8.0
		1	2.54 – 6.0
Distance from water bodies	15 %	5	Depending on distance from:
		4	Sea shore (D_{sea}): 0 – 10 km – 20 km
		3	Lakes (D_{lake}) 0 – 2 km – 4 km
		2	Rivers (D_{river}>) 0 – 1,5 km – 3 km
		1	
Land use & cover	10 %	5	Narrow vegetation, pastures (Codes: 221, 333)
		3	Cultivated land, mixed vegetation, open fields (Codes: 221, 231, 243, 321)
		1	Wetlands, forests, artificial surfaces, water bodies (Codes: 112, 311, 411, 511)

The reason for weighting is to reflect the datasets's relative importance. For example, the elevation layer was weighted with higher importance concerning the influence on frost hazard. Five classes were used (from 1 to 5) for each factor, corresponding to very low, low, medium, high and very high hazard importance as far as frost hazard is concerned. All parameters, classes and their relative weights are presented in Table 3.

For the selection of classes' thresholds in each thematic map the statistical GIS process of Natural Breaks was used (Simpson & Human 2008; Sunbury 2013). It is the most common method for grouping data in GIS. The Natural Breaks method is commonly used to create natural groupings of data, while stressing the differences between their values (Mitchell 1999).

Fig. 5 presents the reclassification of all the related factors. Dark areas indicate higher frost risk whereas in lighter coloured areas there is relative safety from frost.

Finally, all the layers which can be seen in Fig. 5 were overlaid with their relative weight factors, which were assigned based on scientific literature research. Then the map produced was reclassified in 5 frost hazard zones with the Natural Breaks method and the final frost hazard map was produced (see Fig. 8).

The frost hazard parameter combination that was used is as follows:

Frost Hazard $= 0.3 \times$ elevation $+ 0.10 \times$ aspect $+ 0.10 \times$ slope $+ 0.15 \times$ CTI $+$ $0.15 \times$ dist. water $+ 0.10 \times$ curvature $+ 0.10 \times$ landuse   (Eq. 1)

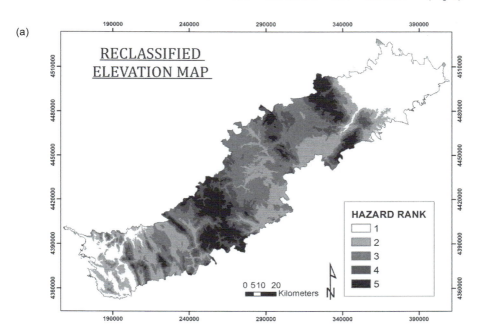

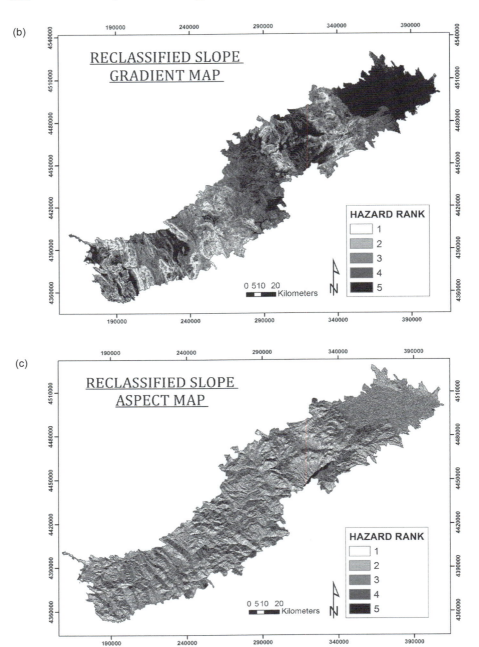

(d)

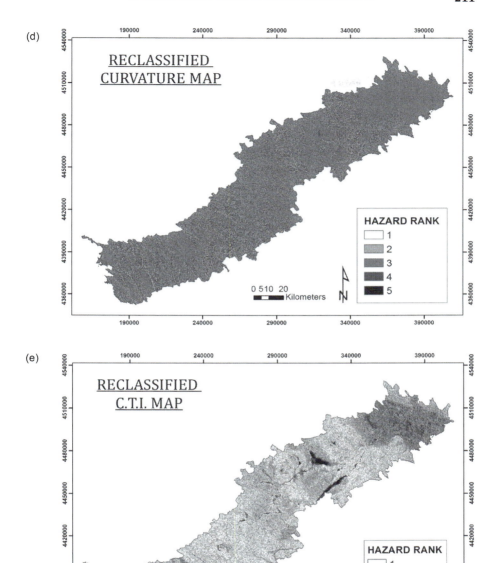

(e)

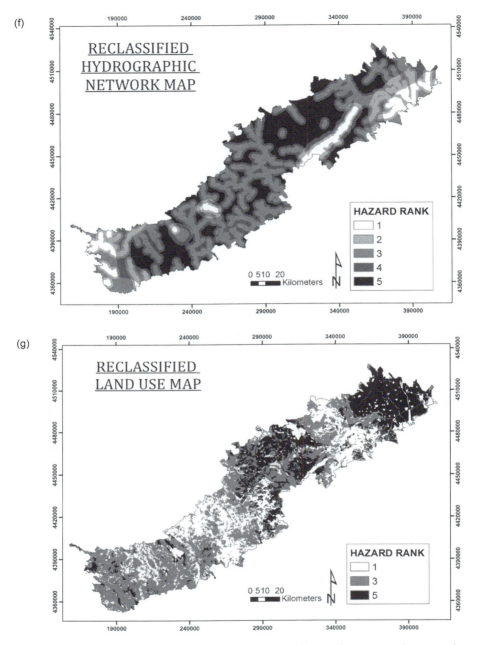

Figure 5 Reclassified frost hazard parameter maps of the study area (a) elevation, (b) slope gradient, (c) slope aspect, (d) curvature, (e) C.T.I, (f) distance from hydrographic network and (g) land use (Projection System: Greek Grid).

Development of Frost Frequency Map

A major part of this study included the use of MODIS satellite data from which the land surface temperature was obtained and used as an input parameter for the development of frost frequency map. This map was produced in order to analyze the frequency of frost and evaluate the results of the frost hazard model.

Firstly, the MODIS operationally distributed products of night temperature were collected and pre-processed. In order to create the frost frequency map, data from 4 years (winter periods) were selected and analyzed, i.e. years 2004, 2006, 2009 and 2010. The selection of the years was based on the analysis of the recorded ground observations concerning frost damages in crops for the decade 2000-2010 by the Greek Agricultural Insurance Organization (ELGA).

A total of 640 MODIS LST images were processed in two stages, as presented in Fig. 6.

The first processing stage concerns the selection of areas where land surface temperature was below 0°C, therefore indicating frost conditions and it included the following steps:

- The images were processed with the Band Math Tool, using the equation:

$$Ts < 273°K \quad \text{(which equals to 0°C)} \tag{Eq. 2}$$

  where Ts represents land surface temperature.
- 640 images were produced that contained pixels with value '0' or '1'. Values of '1' corresponded to pixels where frost was detected and '0' values to pixels which recorded temperature above 0°C.
- The 640 images were then stacked and summed up in one composite image (Image 1). Its pixels' values in the frost frequency layer varied between 1 and 640 and corresponded to the number of days frost was recorded in that particular pixel, from the total of 640 days that were analyzed.

The objective of the second processing stage was to separate the areas where temperature information was recorded from areas where no data were recovered due to cloud cover or other cause). The following processing steps were accomplised:

- Processing of each image with the Band Math Tool, using the equation:

$$Ts < 0°K \tag{Eq. 3}$$

- The 640 images were comprised of pixels with either values '0' or '1'. The pixels with value '1' represented areas where land surface temperature was recorded and '0' values areas where no data was recovered.
- Then, the 640 images that were produced were stacked and summed up in one composite image (Image 2). Its pixel values corresponded to the number of days that there was information of land surface temperature from the total of 640 days.

Finally, the two images were divided, that is, the frost frequency image (Image 1) was divided from the temperature detection frequency image (Image 2). The final result was the frost frequency map that presents the frost frequency in relation to frequency of temperature detection by MODIS instrument.

In order to use this map to validate the classification of the study area, according to the frost hazard model, a classification was conducted on the frost frequency map. The classification was conducted with the 'Natural Breaks' method. The final result of this process is presented in Fig. 9.

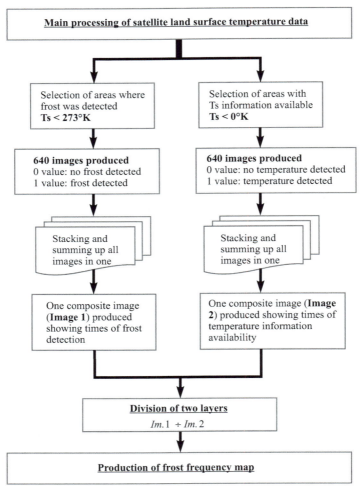

Figure 6 Workflow chart regarding the main processing steps of MODIS land surface temperature data.

Frost Damage on Agricultural Land

In order to study the spatial and temporal distribution of frost risk on the agricultural land of the study area, ground observations by Agricultural Insurance Organization of Greece (ELGA) were processed. Such data that record frost damages are important since they offer an independent constraint and can help validate our model. The data that were processed concerned frost incidents that took place during the decade 2000-2010 and were recorded during the period between 1st of October and 10th of May of each year.

The workflow of this process was composed of the following steps:

(i)   All frost incidents recorded in the 514 sub-municipalities of the study area were selected.

(ii)   Incidents where the inspectors of the organization recorded zero damage were excluded.

(iii) The overall loss rates were computed per sub-municipality. Also, the total losses due to frost per sub-municipality were calculated, expressed both in area of agricultural land and in number of cultivated fruit trees.

(iv)   The sub-municipalities of the study area were classified in five categories depending on the recorded frost damage in fields and trees. The classification grouping was achieved by using the Natural Breaks, GIS method and the final results are presented in the map of Fig. 7.

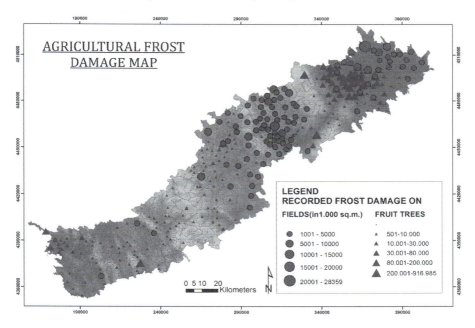

Figure 7 Frost damage in agricultural yields per sub-municipality, as reported by ELGA, for the years 2000-2010 (Projection System: Greek Grid).

The level of reported frost damage in fields per sub-municipality are depicted with red circles, whereas damage recorded as number of damaged fruit trees are presented as red triangles. They are classified into five categories, depending on the recorded amount and rate of frost damage. The size of the symbol is analogous to the level of recorded frost damage.

Frost damage in fields is spread over the whole study area, except for parts of high altitude, where there are no cultivated areas. Damage in fruit trees are reported mainly in the northern part of the study area.

○ RESULTS

Frost Hazard Model

According to the methodology described already, the frost hazard map for the study area was created, after overlaying the different thematic datasets and is presented in Fig. 8.

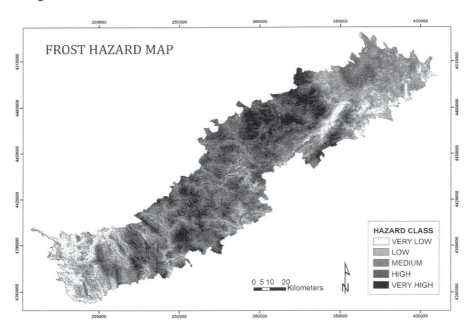

Figure 8 Frost hazard map of the study area produced by the frost hazard model (Projection System: Greek Grid).

The map above highlights areas where the estimated frost hazard level is high (red colored) in which the terrain factors might be more favorable to frost incidents than in other areas (yellow and green colored). Fig. 8 shows significant spatial differences across the study area concerning the estimated susceptibility of frost hazard. Indeed, frost hazard varies significantly over short distances.

High and very high frost hazard zones are situated mostly in the central parts of the study area and especially on higher altitudes, whereas very low and low frost hazard zones are observed in southern and eastern parts. Medium frost hazard zones are scattered throughout the area of interest.

Frost Frequency Map

The result of the process of MODIS surface temperature data, derived following the methodology described earlier is illustrated in Fig. 9. The map of Fig. 9 presents the distribution of frost as it has been captured by the MODIS sensor for the years 2004, 2006, 2009 and 2010.

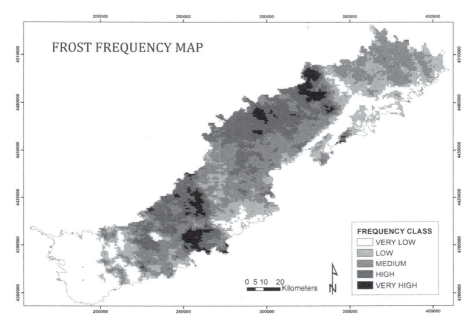

Figure 9 Frost frequency map of the study area developed by MODIS surface temperature data for the years 2004, 2006, 2009, 2010 (Projection System: Greek Grid).

The red areas where the estimated frost frequency is high are concentrated in central parts in the north and south-east. The distribution of frost frequency as presented by the map of Fig. 9 appears to be (in a qualitative sense) quite similar to the results of the frost hazard map (Fig. 8). The latter will be thoroughly analyzed next.

Validation of Frost Hazard Map in Comparison to Frost Frequency Map

The frost hazard map produced from our method (Fig. 8) was compared against the frost frequency map, which has been created using EO-based surface temperature data (Fig. 9). The main goal of this procedure was to calculate the degree of correlation of the two maps, as they were categorized into five zones.

Firstly, the 100 m resolution model results were aggregated in 10 × 10 pixel clusters and the result was compared to the 1 km frost frequency map. The map in Fig. 10 was produced by performing a subtraction of the frost hazard map (Fig. 8) from the frost frequency map (Fig. 9).

The map that was produced (Fig. 10) contained values from '0' to '4' which correspond to the difference of the categorization of the two maps. '0' values represent pixels that were classified in the same category in both the frost hazard map and the frost frequency map, indicating a very high correlation between both maps. Values '1' represent pixels that have a differentiation of one class in the two maps indicating a high correlation between the two maps. Values '2' represent differentiation of two classes, indicating moderate correlation. Values '3' represent differentiation of three classes and indicate low correlation. Values '4' represent pixels that have the highest differentiation between the two maps and correspond to no correlation.

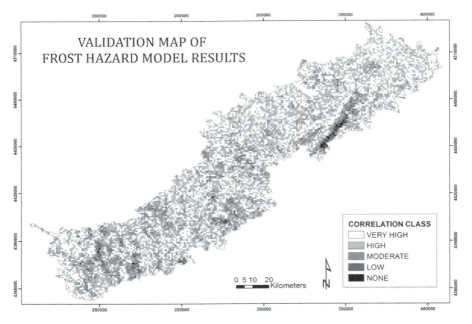

Figure 10 Validation of frost hazard model's results in comparison to the frost frequency map (Projection System: Greek Grid).

According to the validation map (Fig. 10), the two maps are highly correlated, as the correlation class is found to be very high or high in most areas.

The analysis of the validation map shows that the correlation level has been assessed as very high at 36.50% and high at 46.89% of the study area, respectively. As a result, our model correlates well or fairly well with the EO-based surface temperature data over the vast majority (83.39%) of the study area.

It is remarkable that areas of low correlation are concentrated in a zone in the northwestern part. Also, lower correlation is observed in smaller areas northerly to this area (Fig. 11) where the hydrographic network is dense.

TABLE 4 Categorization of the correlation of the frost hazard model results and the frost frequency map.

Correlation class	Percentage
I (Very High)	36.50%
II (High)	46.89%
III (Medium)	14.11%
IV (Low)	2.16%
V (None)	0.34%

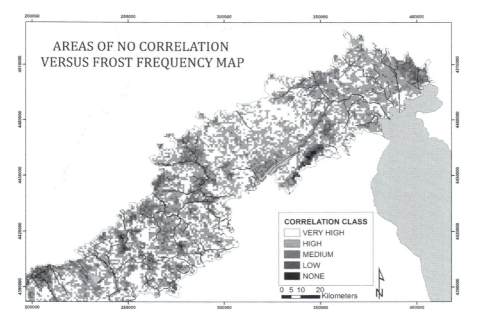

Figure 11 Areas of low correlation levels in the study area (Projection System: Greek Grid).

These results shown above indicate that the parameter of water and especially rivers need to be further analyzed. It seems that rivers provide a protection against frost that depends on the topography of their surroundings. Also, it could be a factor that provides better results in a smaller scale of study, according also to findings of previous studies (Pouteau et al. 2011).

Application of Frost Risk Model on Agricultural Areas

In order to quantify the model's ability to estimate the frost risk levels on agricultural areas, the frost risk model developed in this study was applied on agricultural land. The results obtained were evaluated through a qualitative comparison to ground observations by the Greek Agricultural Insurance Organization (ELGA).

Firstly, the agricultural land (Corine class) of the study area was selected based on the Corine classification of the study area. The land cover data which was used for this process are Version 1 Corine data which corresponds to the year 2000.

The frost risk model was then applied only on the selected areas and the results are presented in the following map (Fig. 12).

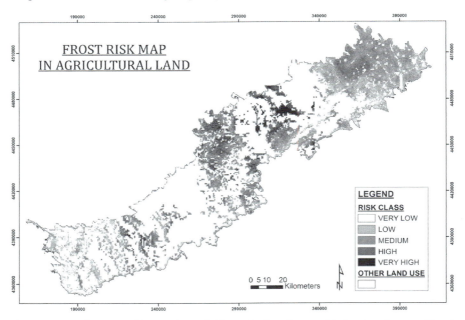

Figure 12 Application of frost risk model in agricultural areas (Projection System: Greek Grid).

As presented in Fig. 12, the agricultural land is mainly located in low frost risk levels (green colored areas). Agricultural land classified in a medium frost risk zone is situated in the northern and central parts of the study area. Agricultural areas exposed in high frost risk (orange and red areas) exist in the northern part of the central sector of the study area.

In the next step of the analysis, the ground observations of frost damage from ELGA (Fig. 7) were overlaid. The map presented in Fig. 13 was produced, which depicts the results of the frost risk model in agricultural land as well as reported frost damage in agricultural yields by ELGA.

Major damages in agricultural yields are recorded in northern and central parts of the study area. Most of those areas are classified by the frost risk model as medium to high frost risk zones. From the map of Fig. 13, adequate agreement is observed in the results of the frost risk model and ground frost observations. It is remarkable that only sparse agricultural frost damages were recorded in areas not classified as agricultural land.

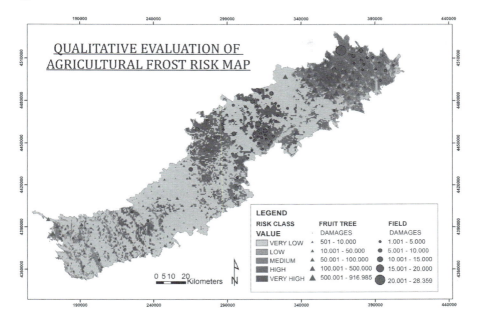

Figure 13 Evaluation of frost risk model results's on agricultural areas in comparison to ground observations of frost damages by ELGA (Projection System: Greek Grid).

○ DISCUSSION & CONCLUDING REMARKS

Mapping of frost hazard in agricultural land has been the main aim of this study. By analyzing and taking into consideration the climatic and geomorphological characteristics of a site, a frost hazard model has been developed.

The frost hazard model was validated in comparison to the distribution of frost hazard in the study area, as it was captured by MODIS Land surface temperature (MOD11A1) data. The results demonstrated a very good correlation between the model and the satellite thermal data. This implies that the spatial distribution of the frost hazard can be adequately mapped through our model.

The frost hazard model was applied on agricultural land and the results were further evaluated through ground observations concerning frost damage in agricultural land and a satisfactory correlation was also found.

One of the strengths of this study is the use of different and independent data sources such as satellite surface temperature data, topographic data and ground observations. This enables validation processes which lead to the extraction of more reliable results.

This study can prove to be useful in various applications. It can be a critical tool in the planning of agricultural policies with significant economic and social benefits. The developed model could be used for civil protection purposes and public transport works and prevention measures.

In addition, the model is a flexible and adaptable tool. With the proper adjustments in the parameterization of frost hazard factors, it could be applied in other areas of the Mediterranean or other climate zones.

It can be integrated with other high resolution data available from both commercial and non-commercial vendors. The data used in the present study although of medium spatial resolution have the advantage of timely and cost free acquisition through the internet. It is also important to mention that the current study has the potential to be developed further or modified. For example, dynamic factors, showing variance with time, could be added so that the models results are more representative of current levels of frost hazard.

The main limitation of the proposed methodology is derived from the uncertainty of the processed data. For example, MODIS land surface temperature data may have significant data gaps (e.g. due to clouds). If these gaps form the predominant part of the database, then the statistical sample could be too short. Such a limited dataset could lead to an unrepresentative frost frequency map that would have to be corrected by combination with ground temperature measurements.

Further studies are required towards the validation of the frost hazard and risk model with ground observations at a wide range of environments. The latter has to be supported by a detailed sensitivity analysis that would test the weight and importance of each modeled parameter.

⭘ ACKNOWLEDGMENTS ━━━━━━━━━━━

The authors are thankful to the Greek Agricultural Insurance Organization (ELGA) for providing data and photographs used in this study. Dr. Petropoulos gratefully acknowledges the financial support provided by the Marie Curie Career Re-Integration Grant "TRANSFORM-EO" project supporting his participation in this research work. He also wishes to thank Mr Matthew North for his assistance with proof reading and language editing.

⭘ REFERENCES ━━━━━━━━━━━

Aber, J.D. and Melillo, J.M. 1991. Terrestrial Ecosystems. Saunders College Publishing, Philadelphia, PA, p. 429.

Almkvist, E., Gustavsson, T. and Borgen, J. 2005. An attempt to define the Road Climate Room. Meteorological Applications 12: 357-370.

Attaway, J.A. 1997. History of Florida citrus freezes. Available online agris.fao.orgdate accessed 01.13.2015.

Beck, E.H., Heim, R. and Hansen, J. 2004. Plant resistance to cold stress: mechanisms and environmental signals triggering frost hardening and dehardening. Journal of Biosciences 29: 449-459.

Berrocal, V.J., Raftery, A.E., Gneiting, T. and Steed, R.C. 2010. Probabilistic weather forecasting for winter road maintenance. Journal of the American Statistical Association 105: 490.

Blennow, K. and Persson, P. 1998. Modelling local-scale frost variations using mobile temperature measurements with GIS. Agricultural and Forest Meteorology 89: 59-71.

Blennow, K. and Lindkvist, L. 2000. Models of low temperature and high irradiance and their application to explaining the risk of seedling mortality. Forest Ecology and Management 135(1): 289-301.

Bolstad, P.V., Bentz, B.J. and Logan, J.A. 1997. Modelling microhabitat temperature for dendroctonus ponderosae (coleoptera: scolytidae). Ecological Modelling 94: 287-297.

Bolstad, P.V., Swank, W. and Vose, J. 1998a. Predicting Southern Appalachian overstory vegetation with digital terrain data. Landscape Ecology 13: 271-283.

Bolstad, P.V., Swift, L., Collins, F. and Regniere, J. 1998b. Measured and predicted air temperatures at basin to regional scales in the southern appalachian mountains. Agricultural and Forest Meteorology 91: 161-176.

Boyer, D.G. 1984. Estimation of daily temperature means using elevation and latitude in mountainous terrain. Water Res. Bull. 20: 583-588.

Caprio, J.M. and Snyder, R.D. 1984. Study to improve winterkill parameters for a winter wheat model. Task 1, a study of the relation between soil temperature at three centimeter depth and air temperature. Final Project report. NASA Contr. NAS 9-16007, p. 76.

Chapman, L., Thornes, J.E. and Bradley, A.V. 2001. Modelling of road surface temperature from a geographical database. Part 2: Numerical Meteorological Applications 8: 421-436.

Chapman, L. and Thornes, J.E. 2006. Road ice prediction using geomatics. Science of Total Environment 3601-3: 68-80.

Chen, G., Peng, M., Huang, R. and Lu, X. 1994. Vegetation characteristics and its distribution of Qilian mountain region (in Chinese with English abstract). Acta Bot. Sinica 36: 63-72.

Chrysoulakis, N., Abrams, M., Feidas, H. and Velianitis, D. 2004. Analysis of ASTER multispectral stereo imagery to produce DEM and land cover databases for Greek islands: The REALDEMS project. Proc. e-Environment Progress and Challenge 411-424.

Chuanyan, Z., Zhongren, N. and Guodong, C. 2005. Methods for modelling of temporal and spatial distribution of air temperature at landscape scale in the southern Qilian mountains, China. Ecological Modelling 189: 209-220.

Cooper, W.C., Young, R.H. and Turrell, F.M. 1964. Microclimate and physiology of citrus: their relation to cold protection. Agricultural Science Review p. 38-50.

Cox, D.L., Larsen, J.K. and Brun, L.L. 1986. Winter survival response of winter wheat: tillage and cultivar selection. Agronomy Journal 78: 795-801.

Domenikiotis, C., Spiliotopoulos, M., Kanellou, E. and Dalezios, N.R. 2003. Classification of NOAA/AVHRR images for mapping of frost affected areas in Thessaly, Central Greece. Proceedings of the International Symposium on GIS and Remote Sensing: Environmental Applications, Volos, Greece p. 25-32.

Drobne, S. and Lisec, A. 2009. Multi-attribute decision analysis in GIS: weighted linear combination and ordered weighted averaging. Nature 4(26): 28.

Dubayah, R. 1994. Modelling a solar radiation topoclimatology for the riogrande river basin. Journal of Vegetation Science 5: 627-640.

Dubayah, R. and Rich, P.M. 1995. Topographic solar radiation models for GIS. International Journal of Geographical Information Systems 9: 405-419.

ELGA – GAIO 2003. Estimation Manual of Frost Damage in Apple Trees. Available online www.elga.grdate accessed 01.13.2015.

Evans, J. 2004. Compound Topographic Index AML". Available on line http://arcscripts.esri. com/details.asp?dbid=11863 date accessed 01.13.2015.

Francois, C., Bosseno, R., Vacher, J.J. and Segiun, B. 1999. Frost risk mapping derived from satellite and surface data over the Bolivian Altiplano. Agricultural and Forest Meteorology 95: 113-137.

FAO 2005. Frost Protection: fundamentals, practice and economics. Rome.

Fridley, J.D. 2009. Downscaling climate over complex terrain: high finescale (<1.000 m) spatial variation of near-ground temperatures in a montane forested landscape (Great Smoky Mountains). Journal of Applied Meteorology and Climatology 48: 1033-1049.

Geiger, R.F., Aron, R.H. and Todhunter, P. 2003. The Climate near the ground. Ed. Rowman and Littlefield Publishers Inc., p. 584, United Kingdom.

Gesch, D., Oimoen, M., Zhang, Z., Meyer, D. and Danielson, J. 2012. Validation of the ASTER Global Digital Elevation Model Version 2 over the conterminous United States. International Archives of the Photogrammetry, Remote Sensing and Spatial Information Sciences, XXXIX (B4).

Gessler, P.E., Chadwick, O.A., Chamran, F., Althouse, L. and Holmes, K. 2000. Modelling soil-landscape and ecosystem properties using terrain attributes. Soil Science Society of America Journal 64: 2046-2056.

Gustavsson, T., Borgen, J. and Erikkson, M. 1998. GIS as a tool for planning new road stretches in respect of climatological factors. Theoretical and Applied Climatology 60: 179-190.

Gustavsson, T. 1999. Thermal mapping – a technique for road climatological studies. Meteological Applications 6(4): 385-394.

Hewitt, K. 1983. Calamity in a technocratic age. Interpretations of Calamity, Allen and Unwin, Boston, MA.

HNMS 1999. Climatic Data of Meteorological Station of HNMS – Issues A and B. Athens, Greece.

Jarvis, C.H. and Stuart, N. 2001. A comparison among strategies for interpolating maximum and minimum daily air temperatures. Part I: The selection of "guiding" topographic and land cover variables. Journal of Applied Meteorology 40: 1060-1074.

Kerdiles, H., Grondona, M., Rodriguez, R. and Seguin, B. 1996. Frost mapping using NOAA AVHRR data in the Papean region, Argentina. Agricultural and Forest Meteorology 79: 157-182.

Kramer, P.J. 1983. Water Relations of Plants. Academic Press, San Diego, CA, p. 489.

Larcher, W. 2001. Ökophysiologie der Pflanzen. Stuttgart: Eugen Ulmer, p. 302.

Leffler, R.J. 1981. Estimating average temperature on Appalachian summits. Journal of Applied Meteorology 20: 637-642.

Lindkvist, L., Gustavsson, T. and Borgen, J. 2000. A frost assessment method for mountainous areas. Agricultural and Forest Meteorology 102: 51-67.

Malczewski, J. 2011. Local weighted linear combination. Transactions in GIS. 15(4): 439-455.

Martsolf, J.D., Gerber, J.F., Chen, E.Y., Jackson, J.L. and Rose, A.J. 1984. What do satellite and other data suggest about past and future Florida freezes. Proc. Fla. State Hort. Soc. 97: 17-21.

Miliaresis, G.C. and Paraschou, C.V. 2011. An evaluation of the accuracy of the ASTER GDEM and the role of stack number: a case study of Nisiros Island, Greece. Remote Sensing Letters 2(2): 127-135.

Mitchell, A. 1999. The ESRI Guide to GIS Analysis. Volume 1: Geographic Patterns and Relationships. Environmental Systems Research Institute, Inc. California.

Oke, T.R. 1987. Boundary Layer Climates. Ed. Routledge, London.

Pouteau, R., Rambal, S., Ratte, J.P., Goge, F., Joffre, R. and Winkel, T. 2011. Downscaling MODIS-derived maps using GIS and boosted regression trees: The case of frost occurrence over the arid Andean highlands of Bolivia. Remote Sensing of Environment 115(1): 117-129.

Radcliffe, J.E. and Lefever, K.R. 1981. Aspect influences on pasture microclimate at Coopers Creek, North Canterbury. New Zealand Journal of Agricultural Research 24: 55-66.

Régnière, J. and Bolstad, P. 1994. Statistical simulation of daily air temperature patterns in eastern north America to forecast seasonal events in insect pest management. Environmental Entomology 23: 1368-1380.

Régnière, J. 1996. A generalized approach to landscape-wide seasonal forecasting in temperature-driven simulation models. Environmental Entomology 25(5): 869-881.

Richards, K. and Baumgarten, M. 2003. Towards Topoclimate Maps of Frost and Frost Risk for Southland, New Zealand. 15th Annual Colloqium of the Spatial Information Research Centre, University of Otago, Dunedin, New Zealand, 01.03.2003.

Russo, J.M., Liebhold, A.M. and Kelley, J.G.W. 1993. Mesoscale weather data as input to a gypsy moth (Lepidoptera: Lymantriidae) phenology model. Journal of Economic Entomology 86: 838-844.

Safanda, J. 1999. Ground surface temperature as a function of slope angle and slope orientation and its effect on the subsurface temperature field. Tectonophysics 306: 367-375.

Sakai, A. and Larcher, W. 1987. Frost survival of plants. Responses and adaptation to freezing stress. Ed. Springer, Berlin.

Shao, J. and Lister, P.J. 1995. The prediction of road surface state and simulation of the shading effect. Boundary Layer Meteorology 73: 411-419.

Simpson, D.M. and Human, R.J. 2008. Large-scale vulnerability assessments for natural hazards. Natural hazard 47(2): 143-155.

Soderstrom, M. and Magnusson, B. 1995. Assessment of local agroclimatological conditions – a methodology. Agricultural and Forest Meteorology 72: 243-260.

Stebelsky, I. 1983. Wheat yields and weather hazards in the Soviet Union. Interpretations of calamity. Allen and Unwin Inc., Boston.

Sunbury, T.M. 2013. The role and challenges of utilizing GIS for public health research and practice. Technology and Innovation 15(2): 91-100.

Turner, M.G. and Gardner, R.H. 1991. Quantitative Methods in Landscape Ecology. Ed. Springer, New York, p. 536.

Waring, R.H. and Schlesinger, W.H. 1985. Forest Ecosystems, Concepts and Management. Academic Press, Orlando, FL, p. 340.

A Statistical Approach for Catchment Calibration Data Selection in Flood Regionalisation

Wan Zurina Wan Jaafar,[1,]* *Dawei Han,*[2]
Prashant K. Srivastava[3,4,5] and *Jia Liu*[6]

ABSTRACT

This study explores the two unsolved problems associated with calibration data for developing a flood regionalisation model. Those problems are: 1) How many calibration catchments should be used? 2) How to select the calibration catchments? Quantity and quality of calibration data are two different entities which could greatly influence the developed model. Their effect on the model performance should be carefully examined and be tested using various catchments. This study therefore explores these two questions through a case study on the median annual maximum flood (QMED) model in the UK. It has been found that the chance of developing a good QMED model decreases significantly when the number of calibration catchments drops below a critical number (e.g., sixty in the case study). However, no significant improvement is achieved if the number of calibration catchments is above it. This number could be used as a threshold for choosing randomly selected calibration catchments. Across a broad range of calibration catchment numbers, there are good and poor calibrated

[1] Department of Civil Engineering, Faculty of Engineering, University of Malaya, 50603 Kuala Lumpur, Malaysia.
[2] Water and Environmental Management Research Centre, Department of Civil Engineering, University of Bristol, Bristol, UK.
[3] Hydrological Sciences, NASA Goddard Space Flight Center, Greenbelt, Maryland, USA.
[4] Earth System Science Interdisciplinary Center, University of Maryland, Maryland, USA.
[5] Institute of Environment and Sustainable Development, Banaras Hindu University, Varanasi, India.
[6] State Key Laboratory of Simulation and Regulation of Water Cycle in River Basin, China Institute of Water Resources and Hydropower Research, Beijing, China.
* Corresponding author: wzurina@um.edu.my

models regardless of calibration catchment numbers. High quality models could be developed from a small number of calibration catchments and while poor models could be developed from a large number of calibration catchments. This indicates that the number of calibration catchments may not be the dominating factor for developing a high quality regionalisation model. Instead, the information content could be more important. The study has demonstrated that the standard deviation values between the best and poorest groups are distinctive and could be used in choosing appropriate calibration catchments.

KEYWORDS: index flood, calibration data selection, flood regionalisation, information content, resampling method, standard deviation.

○ INTRODUCTION

Regional flood frequency analysis is a commonly used method for estimating extreme floods at sites where little or no data are available (Dalrymple 1960; Stedinger and Tasker 1986; Boes et al. 1989; Burn 1990; Rosbjerg and Madsen 1995; GREHYS 1996a; GREHYS 1996b; Pandey and Nguyen 1999; Ouarda et al. 2000; Ouarda et al. 2001; Ouarda et al. 2006; Cunderlick and Burn 2002; Shu and Burn 2004a; Shu and Burn 2004b; Ouarda et al. 2008). Regionalisation is a method that makes use of pooled information from hydrologically similar locations in order to increase the reliability of the flood estimates. Two broad categories of regionalisation methods have been widely used: the index flood approach (Dalrymple 1960) and the multiple regression approach (Benson 1960). The index flood method expresses the flood frequency curve at any site in a homogeneous region as a product of an index flood and a dimensionless regional frequency growth curve. The index flood is usually set equal to the mean or median annual maximum flood. Most studies focus on using the multiple regressions with catchment attributes in predicting flood quantiles or index flood for ungauged catchments (Canuti and Moisello 1982; Acreman 1985; Mimikou and Gordios 1989; Reimers 1990; FEH 1999).

Selection of appropriate catchments for regionalisation model calibration is crucial for the index flood model and flood quantiles at ungauged catchments (Wan Jaafar et al. 2011; Wan Jaafar and Han 2012). However, there are no reported studies on the two unsolved problems in the flood regionalisation model development: 1) How many calibration catchments should be used? 2) How to select the calibration catchments? These questions have been raised by Wan Jaafar and Han (2012) in their article related to the uncertainty in index flood modelling. The authors explored the impact of calibration catchment numbers on the index flood model performance using 182 catchments in England. The statistical resampling method was applied and the outcome indicated that the number of calibration catchments was only one of the factors influencing the model performance. Generally, the intuition of a modeller is that the more calibration data are used, the better the performance of the developed model. Although various numbers of calibration catchments were used by previous studies (as reviewed by Wan Jaafar and Han (2012)) in developing index flood models, there are no consensus on how many catchments should be used and how to choose them.

It is generally accepted that the reliability of the model being developed is highly dependent on the adequacy of the calibration procedure employed. Nowadays, the automatic calibration procedure has been widely used to calibrate the model as this method is much more convenient and faster. The automatic calibration procedure consists of three elements: objective function, optimization algorithm and calibration data. Automatic model calibration has been the main focus of hydrologists in solving hydrologic models in order to produce a better calibrated model. Model parameter estimation using the automatic calibration procedure has been gaining much attention in order to provide more realistic parameter estimates and more reliable forecast. This approach has been discussed extensively in the literature (Sorooshian and Dracup 1980; Sorooshian 1981; Sorooshian et al. 1982, 1983; Kuczera 1983; Gupta et al. 1998). It has been proved that the success of any calibration process is highly dependent on the characteristics (quantity and quality) of the data used. It has often been suggested that the calibration data should be as representative of the various phenomena experienced by the catchment as possible. Research has proved that the information content of the data is far more important than the amount used for model calibration (Kuczera 1982; Sorooshian et al. 1983; Gupta and Sorooshian 1985; Yapo et al. 1996). In this study, the information content is regarded as how informative the selected calibration catchments are in representing all the catchments in a region. Selection of appropriate calibration data has been carried out in a latest study by Liu and Han (2010) for the rainfall-runoff modelling. This study has proposed some indices for the selection of calibration data with adequate lengths and appropriate durations by examining the spectral properties (i.e., in terms of energy distribution in frequency domain) of data sequences before the calibration work. With the validation data determined beforehand, the similarity assumption was applied to find a set of calibration data relevant to the validation data. The more similar the calibration data is to the validation, the better the model performance would be. The similarity between the validation and calibration data was examined using the flow-duration curve, Fourier transform and wavelet analysis. Useful indices such as information cost function and an entropy-like function were used to evaluate the results of the three methods. This study has found that information content of the calibration data was more important than the data length. Shorter data length may provide more useful information than a longer data series.

In this study, further exploration on calibration catchments that are likely to produce a high quality model is investigated. Besides, this study examines an appropriate catchment indicator (i.e., an index representing the information content of a group of data) for choosing an appropriate group of calibration catchments provided that a group of validation catchments are determined in advance. The ultimate aim of this study is to explore and answer the two aforementioned questions which were unsolved in the previous study by Wan Jaafar and Han (2012).

○ CATCHMENT DESCRIPTORS AND STUDY AREAS

There are 182 catchments chosen across England and out of them, 137 are used for calibration and the remaining 45 for validation. Input variables (catchment characteristics) used in this study are obtained from the Flood Estimation Handbook (the FEH CD-ROM 2.0) and Environment Agency's HiFlow website (http://www.environment-agency.gov.uk/hiflows/97503.aspx). In order to focus the study on relatively more homogenous data, several conditions have been set on the catchments used in this study: 1) in England only; 2) AREA less than 1000 km^2; 3) QMED value indicated as valid and the relevant flow records as still 'open' in the HiFlows online database; 4) catchments with URBEXT < 0.025 (i.e., mostly rural); 5) QMED less than 50 cumecs. After the screening, only 182 catchments satisfy the set conditions and they are shown in Fig. 1. Among them, 137 (75%) of the catchments are devoted for calibration and the other 45 (25%) catchments are for validation. The validation catchments are randomly chosen to be representative based on the observed QMED value. Detailed explanation on the input variable selection is not elaborated in this paper hence; interested readers should refer to Wan Jaafar and Han (2012).

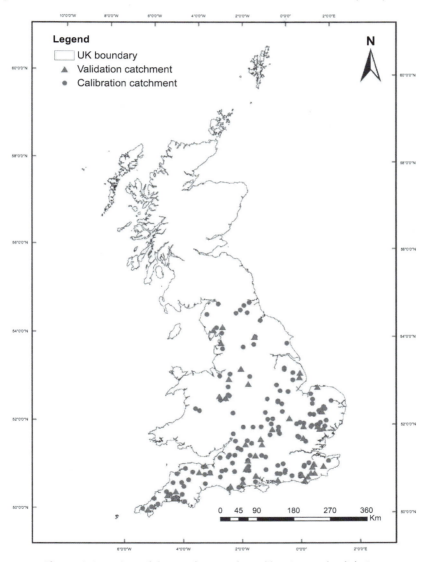

Figure 1 Location of the catchments for calibration and validation.

○ METHODOLOGY

In order to answer the key questions in this study, an index flood model based on median annual maximum flood (QMED) needs to be firstly developed. This model is constructed using an input variable selection technique from the data mining field where the best model is chosen on the basis of the generalisability to avoid choosing an overfitted/underfitted model. Generalisability explains whether a model is a good predictor of future observations (Myung et al. 2009). Comparison between the alternative developed model and the Flood Estimation Handbook (FEH) model used by practitioners in the UK is then examined. Further explanation is not described here because it has been detailed by Wan Jaafar and Han (2012). Once this model is created, performance of different numbers of calibration catchments is studied by employing the statistical resampling technique to search for the minimum number of calibration catchments for high quality model development. Based on this result, a suitable indicator for choosing a group of calibration catchments is investigated provided that the validation group of catchments is determined in advance.

The resampling without replacement method is adopted to ensure that no repetition occurs in the each run. This method is used to select various numbers of calibration catchments from 15, 20, 30, 40, 50, 60, 70, 80, 90, 100, 110, 120 and 130 out of 137. By using the alternative QMED model, the calibration process is performed 1000 times and the resulting models are validated to the 45 data that have been set aside. A minimum number of catchment is decided by simply examining the model performance on validation results presented by the box plot graph. The 1000 model performance results indicated by the coefficient of determination (R^2) is used as a performance evaluation criterion

$$R^2 = 1 - \left(\sum_{i=1}^{N} (\text{QMED}_i^{\text{observed}} - \text{QMED}_i^{\text{modelled}})^2 / \sum_{i=1}^{N} (\text{QMED}_i^{\text{observed}} - \overline{\text{QMED}}^{\text{observed}})^2 \right)$$

(1)

The coefficient of determination value provides useful information about how well a model is able to estimate the future QMED and its value ranges from 0 to 1. Performance of the model resulting from the 1000 simulation runs is plotted against different numbers of calibration catchments. The 1000 validation results are rearranged in magnitude order so that the calibration catchments with the poorest or best result could be identified. Information content of the calibration and validation groups is linked with the performance from the developed model and by examining the data distribution between these two groups; the standard deviation is decided as an indicator.

$$\text{STD} = \sqrt{\frac{1}{N-1} \sum_{i=1}^{N} (x_i - \overline{x})^2}$$

(2)

where x_1, x_2, ..., x_N are the observed values of the sample items, $\overline{x}$ is the mean value of these observations and N is the sample size. The standard deviation value is computed for the selected groups of calibration catchments (100 catchments in each group) chosen from the highest and the lowest validation results. The standard deviation values are computed for every selected calibration group and these values are then averaged. Consequently, the computed average value is compared to the standard deviation value of the validation group.

$$\text{PBIAS} = \sum_{i=1}^{N} (\text{STD}_i^{\text{sampling}} - \text{STD}_i^{\text{validation}}) / \sum_{i=1}^{N} \text{STD}_i^{\text{validation}} \times 100\% \qquad (3)$$

Percentage difference of the standard deviation value between the poorest and best calibration catchment groups with respect to the validation group is measured by calculating the percentage bias (PBIAS) defined by Eq (3). The STD in this equation represents the standard deviation value of either sampling or validation group and N is the number of groups. This calculation is repeated for the 15, 20, 30, 40, 50 and 60 calibration catchments. The standard deviation values obtained from the different number of calibration catchments are finally averaged. The final averaged value is subsequently used as a parameter setting value for the next analysis where the calibration and validation processes are to be repeated.

At this stage, the resampling technique is repeated for 10000 times with a variable constraint imposed during the calibration catchments selection. This variable constraint is the computed average value of the standard deviation being imposed to each variable of the newly-developed model. Out of 10000 runs, only groups of the selected calibration catchments that satisfy the variable setting values are chosen for validation. Finally model performance of the newly-selected calibration catchments on the basis of the variable setting values is then compared to the previous validation result. This comparison is done for the 15, 20, 30, 40, 50 and 60 calibration catchments by applying the similar variable setting values.

◯ RESULTS

The best QMED model from the aforementioned method is shown by Eq (4) consisting of seven variables (i.e., AREA, SPRHOST, DPSBAR, FARL, SAAR, BFIHOST, PROPWET) (Wan Jaafar and Han (2011)). These variables are defined as follow: QMED (median annual maximum flood (m^3/s)), AREA (area of catchment (km^2)), SPRHOST (standard percentage runoff derived by using the hydrology soil type classification), DPSBAR (mean of all the inter-nodal slopes for the catchment, characterizes the overall steepness (m/km)), FARL (flood attenuation by reservoirs and lakes), SAAR (average annual rainfall (mm) from 1961-1990 (mm)), BFIHOST (base flow index derived by using the hydrology soil type classification) and PROPWET (proportion of time when soil moisture deficit (SMD) was equal to, or below, 6 mm during 1961-1990) (FEH 1999).

$$\text{QMED} = 201.1357 \times (\text{AREA}/1000)^{0.7611} \times (\text{SPRHOST}/100)^{0.6914} \times$$
$$(\text{DPSBAR}/100)^{0.3375} \times \text{FARL}^{3.6624} \times (\text{SAAR}/1000)^{1.0929} \times$$
$$\text{BFIHOST}^{-0.6192} \times \text{PROPWET}^{-0.0371} \qquad (4)$$

By using this model, the subsequent step of resampling analysis is conducted by randomly picking up the calibration catchments and followed by model evaluation. Fig. 2 shows a box plot graph whereas Table 1 indicates statistical analysis of the validation results (R^2) in terms of minimum, maximum, median, mean, lower quartile (Q1) and upper quartile (Q3) based on the 1000 resamples using different numbers of calibration catchments. The maximum values across all the different numbers of calibration catchments seem similar but the minimum values vary significantly especially for the smallest calibration catchment at 15. However, there are only 3.2% of the validation results given by the 15 calibration catchments with R^2 of more than 0.7. The extent of dispersion of the R^2 values given by the different numbers of calibration catchments can be quantified by measuring the differences between the Q3 (75 percentile) and Q1 (25 percentile) of the box plot graph. The minimum number of calibration catchments required for developing a high quality model can be ascertained by observing the box plot in Fig. 2 and graph in Fig. 3. The Q3 (the upper line of the box plot) differs for the different numbers of calibration catchments and the differences between the two consecutive numbers of calibration catchments are measured. The percentage difference of the Q3 is depicted by Fig. 3 and the largest difference is between 15 and 20 calibration catchments. From 60 calibration catchments and above, the Q3 differences are relatively similar and small (less than 1%). Therefore, 60 calibration catchments can be used as a critical point where the minimum number of calibration catchments is set.

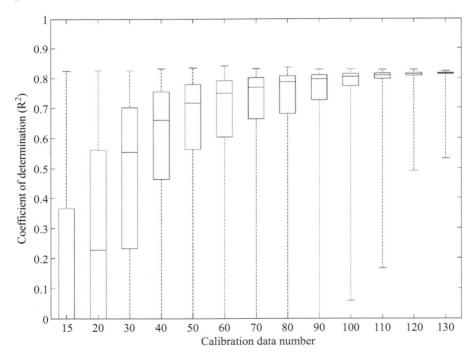

Figure 2 Performance of 1000 validation results.

TABLE 1 Statistical analysis of the R^2 value for validation result.

Data size	Minimum	Maximum	Median	Mean	Q_1	Q_3
15	−8.85E+19	0.824	−0.404	−1.765E+17	−4.055	0.366
20	−1.68E+17	0.826	0.228	−1.676E+14	−1.031	0.561
30	−179.4770	0.826	0.553	−0.070	0.233	0.701
40	−17.100	0.833	0.662	0.454	0.463	0.755
50	−3.090	0.835	0.717	0.597	0.563	0.780
60	−1.079	0.841	0.749	0.652	0.604	0.792
70	−0.772	0.832	0.769	0.693	0.664	0.801
80	−0.398	0.836	0.787	0.719	0.681	0.807
90	−0.195	0.830	0.797	0.742	0.725	0.810
100	0.059	0.830	0.804	0.762	0.773	0.814
110	0.166	0.828	0.810	0.783	0.797	0.816
120	0.491	0.828	0.813	0.797	0.807	0.817
130	0.532	0.823	0.816	0.809	0.813	0.818

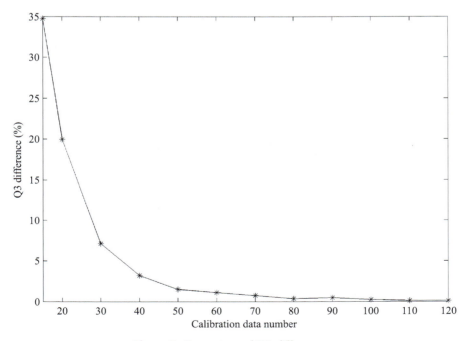

Figure 3 Percentage of Q3 differences.

A further investigation identifies an appropriate indicator that can be used in order to select a group of calibration catchments provided that a group of validation catchments are determined in advance. The information content between the best and poorest groups is explored by examining the spread of each variable. The standard deviation is used as a spread indicator and this value is computed for the entire variables of the QMED model (i.e., AREA, SPRHOST, DPSBAR, FARL, SAAR, BFIHOST, PROPWET). There are two hypotheses: 1) the more spread the variables, the better the developed model should be as the calibration data would be more informative across the data spectrum and 2) the more similar the calibration data to the validation data, the better the model would be developed. Both hypotheses have their own merits and we need numerical experiments to check if they are valid. In doing so, 100 groups of the poorest and best validation results are selected. The standard deviation values are computed for each group and these values are averaged. A similar calculation is repeated for every different number of calibration catchments (i.e., 15, 20, 30, 40, 50 and 60). The calculation is stopped at 60 as this is the minimum number of calibration catchments obtained based on the previous analysis. As discussed previously, the Q3 of the box plot (Fig. 2) exhibits the largest drop starting from 60 until 15 calibration catchments. The standard deviation values for 100 groups are presented in Table 2 with the last column indicating the final average of the standard deviation value. Standard deviation values for the validation group are also measured for the purpose of comparison. For the validation result from the best group, most of the variables except SPRHOST have the average standard deviation values higher than the poorest group and are closer to those of the validation group (Table 2). The best calibration group spread even more than the validation group for the variables SAAR, BFIHOST and PROPWET. For SPRHOST, the spread of the best calibration group is inconsistent as compared to the poorest calibration group and has the final average standard deviation value of 0.02 less than the poorest calibration group. The percentage bias (PBIAS) of the final average standard deviation value of the poorest and best calibration groups with respect to the validation group is presented in Table 3. The negative sign of PBIAS in this table implies that the poorest or best calibration group has less spread than the validation group. All variables of the calibration groups exhibit less spread except for the three variables, i.e. SAAR, BFIHOST and PROPWET. The PBIAS values for all variables are depicted as a bar chart shown in Fig. 4. This figure indicates that all variables of the best calibration group spread more than the poorest calibration group. Variables QMED, AREA, SPRHOST, DPSBAR and FARL show negative PBIAS values because both groups have less spread as compared to the validation group. Nevertheless, the best group of all variables except SPRHOST are more spread as they give the PBIAS value higher than the poorest group. For SPRHOST, the percentage difference of PBIAS between the best and poorest calibration groups is 0.17% which can be neglected and the standard deviation values for those two groups can be considered as similar. The extent of spread of the best and poorest groups can be seen clearly in Fig. 4. In conclusion, the spread characteristic measured by the standard deviation is useful in examining the information content of the calibration data.

TABLE 2 Standard deviation (STD) for best and poorest calibration groups.

Variable	STD for selected calibration data						Average STD
	15	20	30	40	50	60	
QMED							
Best	12.413	12.740	12.703	12.601	12.519	12.637	12.602
Poor	12.255	12.160	12.298	12.377	12.430	12.524	12.341
Validation	13.834	13.834	13.834	13.834	13.834	13.834	13.834
AREA							
Best	111.691	117.806	118.579	112.765	114.796	113.997	114.939
Poor	98.086	105.766	96.024	103.109	104.563	108.419	102.661
Validation	133.620	133.620	133.620	133.620	133.620	133.620	133.620
SPRHOST							
Best	12.008	11.684	11.760	11.663	11.577	11.662	11.726
Poor	11.718	11.264	11.956	11.841	11.830	11.865	11.746
Validation	11.890	11.890	11.890	11.890	11.890	11.890	11.890
DPSBAR							
Best	48.083	46.896	45.962	46.657	46.264	45.501	46.561
Poor	37.651	37.537	36.504	37.565	36.938	37.320	37.253
Validation	47.459	47.459	47.459	47.459	47.459	47.459	47.459
FARL							
Best	0.044	0.042	0.041	0.041	0.041	0.042	0.042
Poor	0.039	0.035	0.039	0.039	0.038	0.038	0.038
Validation	0.043	0.043	0.043	0.043	0.043	0.043	0.043
SAAR							
Best	375.359	386.07	370.8747	365.9213	367.019	366.993	372.039
Poor	329.741	315.34	316.54	337.04	327.47	335.163	326.882
Validation	344.479	344.479	344.479	344.479	344.479	344.479	344.479
BFIHOST							
Best	0.178	0.174	0.174	0.172	0.172	0.172	0.174
Poor	0.139	0.133	0.142	0.141	0.142	0.142	0.140
Valid	0.169	0.169	0.169	0.169	0.169	0.169	0.169
PROPWET							
Best	0.117	0.116	0.111	0.111	0.110	0.109	0.112
Poor	0.098	0.102	0.100	0.104	0.103	0.103	0.102
validation	0.104	0.104	0.104	0.104	0.104	0.104	0.104

TABLE 3 Percentage bias of average STD between the validation data and the poorest and best groups.

Variable	Average STD	PBIAS	% difference
QMED			
Best	12.602	−8.906	1.890
Poor	12.341	−10.796	
Validation	13.834		
AREA			
Best	114.939	−13.981	9.189
Poor	102.661	−23.170	
Validation	133.620		
SPRHOST			
Best	11.726	−1.381	−0.168
Poor	11.746	−1.213	
Validation	11.890		
DPSBAR			
Best	46.561	−1.894	19.613
Poor	37.253	−21.506	
Validation	47.459		
FARL			
Best	0.042	−3.371	8.851
Poor	0.038	−12.222	
Validation	0.043		
SAAR			
Best	372.039	8.001	13.109
Poor	326.882	−5.108	
Validation	344.479		
BFIHOST			
Best	0.174	2.759	20.022
Poor	0.140	−17.263	
Validation	0.169		
PROPWET			
Best	0.112	7.939	10.240
Poor	0.102	−2.301	
Validation	0.104		

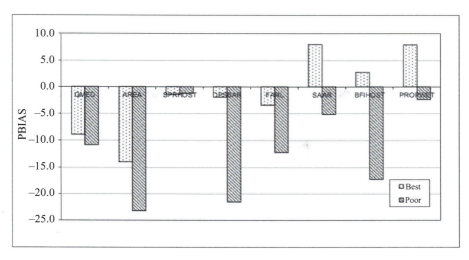

Figure 4 Percentage bias of the average STD value between the validation data and the best & poorest groups.

In the follow-up step, the standard deviation indicator is applied for the purpose of choosing calibration catchments. The final averages of the standard deviation values are utilised as parameter setting values being imposed during the calibration catchments selection. Only groups of calibration catchments that satisfy the parameter setting values are to be chosen. The number of resampling has been increased up to 10,000 to search for the catchment groups that satisfy the criteria. The newly-selected calibration catchments based on the parameter setting values are then validated. This process is repeated for the 15, 20, 30, 40, 50 and 60 calibration catchments using similar parameter setting values. The parameter setting value for each variable is presented in Table 4. The standard deviation values of the best and poorest groups in column two and three are the final average values of the standard deviation taken from Table 3. A good calibration catchment group should exhibit a standard deviation value closer or more than the validation group. Therefore, the standard deviation value from the best group is used as a parameter setting. The newly-selected calibration catchments based on the parameter setting values are validated. The dispersion of model performance (R^2) on the 1000 validation results are presented in the box plot graph which indicates how large the difference is between the upper and lower quartile of the box plot. Comparison of the calibration catchments selected without parameter setting and with parameter setting for the 15, 20, 30, 40, 50 and 60 calibration catchments is depicted in Fig. 5. Obviously the calibration catchments selected with the parameter setting have achieved some great improvement over the randomly selected ones. The dispersion extent of the model performance for the 15 calibration catchments is much less and the median quartile has improved to 0.53. The dispersion extent of 60 to 15 calibration catchments is narrowed as compared to the dispersion extent of the randomly selected calibration catchments without parameter setting. Table 5 provides further details of the quartile values across different numbers of calibration catchments. All the quartile values of the calibration catchments selected with the parameter setting have shown significant improvement especially for the lowest quartile (Q1).

TABLE 4 Parameter setting of the standard deviation value.

	Best group	Poor group	Validation data	Variable setting
QMED	12.602	12.341	13.834	QMED >= 12.602
AREA	114.939	102.661	133.62	AREA >= 114.939
SPRHOST	11.726	11.746	11.890	SPRHOST >= 11.726
DPSBAR	46.561	37.253	47.459	DPSBAR >= 46.561
FARL	0.042	0.038	0.043	FARL >= 0.042
SAAR	372.039	326.882	344.479	SAAR >= 372.039
BFIHOST	0.174	0.14	0.169	BFIHOST >= 0.174
PROPWET	0.112	0.102	0.104	PROPWET >= 0.112

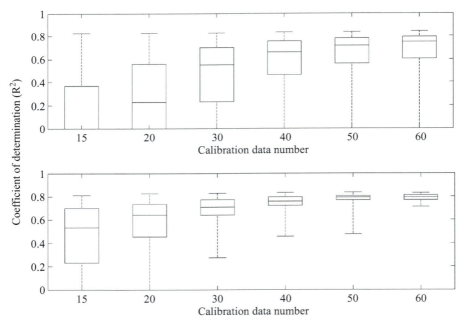

Figure 5 Model performance of the calibration catchments selected without parameter setting (the top figure) and with parameter setting (the bottom figure).

TABLE 5 Spreadability of the model performance in terms of quartile value for the calibration catchments selected with or without parameter setting value.

Calibration Number	Q1 (25 percentile)		Q2 (50 percentile)		Q3 (75 percentile)	
	Without	With	Without	With	Without	With
15	−4.055	0.232	−0.404	0.535	0.366	0.703
20	−1.031	0.455	0.228	0.643	0.561	0.736
30	0.233	0.641	0.553	0.710	0.701	0.773
40	0.463	0.722	0.662	0.757	0.755	0.796
50	0.563	0.767	0.717	0.791	0.780	0.805
60	0.604	0.770	0.749	0.794	0.792	0.811

○ DISCUSSION AND CONCLUSIONS

In flood regionalisation studies, model uncertainty due to the calibration catchments selection has been largely ignored by the hydrological community. Issues related to the minimum number of calibration catchments and types of catchment suitable for model development are not yet explored so far. Using the index flood regionalisation model (QMED), this study explores these issues. It is common for a modeller to use as many calibration catchments as possible since it is believed a good model should be developed with a large number of calibration data.

Information content in calibration catchments is found to be a more crucial factor than the number of calibrations. This information may be quantified provided that the validation data is determined in advance. Careful selection of the calibration catchments with the similarity of information content close to the validation catchments would result in a high quality model. The advantages of selecting calibration catchments on the basis of information content are: 1) significant improvement of the calibrated model is likely achievable; 2) it requires a small number of catchments and this could be beneficial if the data availability is a major problem and 3) it would save time and effort because the model calibration process would be carried out on a small number of calibration catchments. This is achievable by applying an appropriate indicator in selecting calibration catchments. This study has found the standard deviation could quantify an extent of similarity between the two comparison groups and is a useful measure for this purpose. The results have shown that significant improvement is achieved after imposing a constraint of the standard deviation to the variables.

To conclude, this study has been designed to tackle two important questions in developing flood regionalisation models: 1) How many calibration catchments should be used? 2) How to select the calibration catchments? Through a resampling technique, the selection of different numbers of calibration catchments has been performed and this study has found that 60 catchments as a critical point for model calibration. No significant improvement is achieved if the number of calibration catchments is above 60. But the chance of developing a good model decreases significantly when the number of calibration catchments drops below 60. This number could be used as a threshold for choosing randomly selected calibration catchments. Although such a number could vary for different climatic and geographical locations, the presented methodology should still be applicable. Further investigation reveals that across a broad range of catchment calibration numbers, there are good and poor calibrated models regardless of the calibration data numbers. High quality models could still be developed from a small number of calibration catchments and there are also poor models even with a large number of calibration catchments. This indicates that the number of calibration catchments is not the dominating factor for developing a high quality model. Instead, the information content of the calibration catchments is more important. A small group of catchments with large spread of catchment characteristics would be more useful than that of a large group of catchments with less spread. We have explored the standard deviation measure to represent the extent of spread of the catchment characteristics. The results have demonstrated that the standard deviation values between the best and poorest groups are indeed distinctive. If we use the standard deviation to choose the calibration catchments, it is possible to build a more accurate model with a smaller number of calibration catchments. Such knowledge is very important for locations where the catchment data availability is limited. Statistical resampling technique seeking the large standard deviation will tend to select catchments with varied sizes (i.e., to increase the spread), therefore there won't be biases toward smaller or larger drainage areas. It should be noted that this study is the first of its kind in the field of hydrological regionalisation model development and further exploration of the proposed methodology is needed by the hydrological community to improve the methodology and fill in the knowledge gaps in different regional conditions.

○ REFERENCES

Acreman, M.C. 1985. Predicting the mean annual flood from basin characteristics in Scotland. Journal of Hydrological Sciences 30(1): 37-50.

Benson, M.A. 1960. Characteristics of frequency curves based on a theoretical 1000-year record. *In*: Dalrymple, T. (ed.). Flood-Frequency Analysis. U.S. Geological Survey Water-Supply Paper 1543-A, 51-74.

Boes, D.C., Heo, J. and Salas, J.D. 1989. Regional flood quantile estimation for a Weibull model. Water Resource Research 25(5): 979-990.

Burn, D.H. 1990. Evaluation of regional flood frequency analysis with a region of influence approach. Water Resource Research 26(10): 2257-2265.

Canuti, P. and Moisello, U. 1982. Relationship between the yearly maxima of peak and daily discharge for some basins in Tuscany. Journal of Hydrological Sciences 27(2): 111-128.

Cunderlik, J.M. and Burn, D.H. 2002. Analysis of the linkage between rain and flood regime and its application to regional flood frequency estimation. Journal of Hydrology 261(1-4): 115-131.

Dalrymple, T. 1960. Flood frequency analysis. Water Supply Paper 1543-A, U.S. Geological Survey, Washington, DC.

FEH. 1999. Flood Estimation Handbook. Centre for Ecology and Hydrology, 5 volumes, Wallingford, UK.

GREHYS (Groupe De Recherche En Hydrologie Statistique). 1996a. Presentation and review of some methods for regional flood frequency analysis. Journal of Hydrology 186(1-4): 63-84.

GREHYS (Groupe De Recherche En Hydrologie Statistique), 1996b. Intercomparison of flood frequency procedures for Canadian rivers. Journal of Hydrology 186(1-4): 85-103.

Gupta, H.V., Sorooshian, S. and Yapo, P.O. 1998. Towards improved calibration of hydrologic models: multiple and noncommensurable measures of information. Water Resource Research 34(4): 751-763.

Gupta, V.K. and Sorooshian, S. 1985. The relationship between data and the precision of parameters estimates of hydrologic models. Journal of Hydrology 81(1-2): 57-77.

Kuczera, G. 1982. On the relationship of the reliability of parameter estimates and hydrologic time series data used in calibration. Water Resource Research 18(1): 146-154.

Kuczera, G. 1983. Improved parameter inference in catchment models: 1. evaluating parameter uncertainty. Water Resource Research 19(5): 1151-1162.

Liu, J. and Han, D. 2010. Indices for calibration data selection of the rainfall-runoff model. Water Resource Research 46(4): 1-17.

Mimikou, M. and Gordios, J. 1989. Predicting the mean annual flood and flood quantiles for ungauged catchments in Greece. Journal of Hydrological Sciences 34(2): 169-184.

Myung, J., Tang, Y. and Pitt, M.A. 2009. Evaluation and comparison of computational models. Methods in Enzymology 454: 287-304.

Ouarda, T.B.M.J., Haché, M., Bruneau, P. and Bobée, B. 2000. Regional flood peak and volume estimation in northern Canadian basin. Journal of Cold Regions Engineering 14(4): 176-191.

Ouarda, T.B.M.J., Girard, C., Cavadias, G.S. and Bobée, B. 2001. Regional flood frequency estimation with canonical correlation analysis. Journal of Hydrology 254(1-4): 157-173.

Ouarda, T.B.M.J., Cunderlik, J., St-Hilaire, A., Barbet, M., Bruneau, P. and Bobée, B. 2006. Data-based comparison of seasonality-based regional flood frequency methods. Journal of Hydrology 330(1-2): 329-339.

Ouarda, T.B.M.J., Ba, K.M., Diaz-Delgado, C., Carsteanu, A., Chokmani, K., Gingras, H., Quentin, E., Trujillo, E. and Bobée, B. 2008. Intercomparison of regional flood frequency estimation methods at ungauged sites for a Mexican case study. Journal of Hydrology 348(1-2): 40-58.

Pandey, G.R. and Nguyen, V-T-V. 1999. A comparative study of regression based methods in regional flood frequency analysis. Journal of Hydrology 225(1-2): 92-101.

Reimers, W. 1990. Estimating hydrological parameters from basin characteristics for large semiarid catchments. Ljubljana Symposium on Regionalisation in Hydrology. IAHS Publ. no. 191.

Rosbjerg, D. and Madsen, H. 1995. Uncertainty measures of regional flood frequency estimators. Journal of Hydrology 167(1-4): 209-224.

Shu, C. and Burn, D.H. 2004a. Artificial neural network ensembles and their application in pooled flood frequency analysis. Water Resources Research 40(9): W09301.

Shu, C. and Burn, D.H. 2004b. Homogeneous pooling group delineation for flood frequency analysis using a fuzzy expert system with genetic enhancement. Journal of Hydrology 291(1-2): 132-149.

Sorooshian, S. and Dracup, J.A. 1980. Stochastic parameter estimation procedures for hydrologic rainfall-runoff models: correlated and heteroscedastic error cases. Water Resource Research 16(2): 430-442.

Sorooshian, S. 1981. Parameter estimation of rainfall-runoff models with heteroscedastic streamflow errors: the noninformative data case. Journal of Hydrology 52(1-2): 127-138.

Sorooshian, S., Gupta, V.K. and Fulton, J.L. 1982. Parameter estimation of conceptual rainfall-runoff models assuming autocorrelated data errors: a case study, in Statistical analysis of Rainfall and Runoff, Proceedings of the International Symposium on Rainfall Runoff Modelling, Mississippi, edited by Singh, V.P., WRP, Littleton, Colorado, 491-504.

Sorooshian, S., Gupta, V.K. and Fulton, J.L. 1983. Evaluation of maximum likelihood parameter estimation techniques for conceptual rainfall-runoff models: influence of calibration data variability and length on model credibility. Water Resource Research 19(1): 251-259.

Stedinger, J.R. and Tasker, G.D. 1986. Regional hydrologic analysis, 2, model-error estimators, estimation of sigma and log-pearson type 3 distributions. Water Resource Research 22(10): 1487-1499.

Yapo, P.O., Gupta, H.V. and Sorooshian, S. 1996. Automatic calibration of conceptual rainfall-runoff models: sensitivity to calibration data. Journal of Hydrology 181(1-4): 23-48.

Wan Jaafar, W.Z., Liu, J. and Han, D. 2011. Input variable selection for median flood regionalisation, Water Resources Research 47(7): 1-18.

Wan Jaafar, W.Z. and Han, D. 2012. Uncertainty in index flood modelling due to calibration data sizes, Hydrological Processes 26(2): 189-201.

15
CHAPTER

Prediction of Caspian Sea Level Fluctuations Using Artificial Intelligence

Moslem Imani, Rey-Jer You* and Chung-Yen Kuo

ABSTRACT

An analysis and accurate prediction of the sea level fluctuations in Caspian Sea is always important because it potentially affects the natural processes occurring in the basin and influences the infrastructure built along the coastlines. In this article, different approaches in analysis and forecasting of Caspian Sea level anomalies derived satellite altimetry are presented. Compared with the conventional linear regression methods, such as the routine Autoregressive Moving Average (ARMA) models, neural network methodologies and artificial intelligence approaches are the more powerful tools in providing reliable results of the short-term Caspian Sea level anomaly prediction according to our study. Based on the analysis of the minimum Root Mean Square Error and the maximum coefficient of determination, the Support Vector Machine (SVM) shows the best performance in predicting the Caspian Sea level anomalies. The excellent methods and models for the Caspian Sea level analysis included in this article can be employed to monitor water level changes in other water bodies whose time series of water levels have the stochastic behavior in the future.

KEYWORDS: Sea level, ARMA, Artificial Intelligence, Caspian Sea.

○ INTRODUCTION

Nowadays, sea level change is one of the consequences of climate change and is an important issue for both societies and environments. The sea level changes vary

Department of Geomatics, National Cheng Kung University, Tainan City, Taiwan.
* Corresponding author: rjyou@mail.ncku.edu.tw

over a big range of spatial and temporal scales, from seconds to millions of years. Since the changes of sea level are contributed to by lots of factors related to the hydro-meteorological elements, they can be considered as an integral measure of climate change (Milne et al. 2009; Church et al. 2010).

The coastal zone is one of the major areas of human habitation and economic activity (Turner et al. 1996; Sachs et al. 2001; Senapati and Gupta 2014). In 1990, about 1.2 billion of the world's population lived in the near-coastal zone and the population density there was about three times higher than the global average (Small and Nicholls 2003). The sea level rise or lake water level changes will have a potential threat the nature, environment and socio-economic activities, such as flood risk, damage of social and commercial infrastructure, loss of properties and life (Brunel and Sabatier 2009; Singh and Singh 2013; Takagi et al. 2014; Pycroft et al. 2015). A study and prediction of sea level changes is therefore an extremely important issue which may provide sufficient and useful information to prevent natural disasters for life living near water areas.

○ SHORT-TERM VARIABILITY AND LONG-TERM TRENDS

According to a lot of previous researches, sea level records contain a significant decadal variability (e.g. Bindoff and Willebrand 2007; Lebedev and Kostianoy 2008); for example, approximately a 10 mm rise and fall of global mean sea level in the 1997-1998 El Niño-Southern Oscillation (ENSO) event and a temporary 5 mm fall in the 2010-2011 event. Strong evidence shows that the rate of sea level rise increased between the mid-19th and mid-20th centuries (Bindoff and Willebrand 2007). Jevrejeva et al. (2008) also reported that sea level rose by 6 cm during the 19th century and 19 cm in the 20th century. Sea level acceleration up to now is about 0.01 mm/yr^2 since the end of the 18th century (Svetlana et al. 2008). These studies reveal that the short-term variability and long-term trends of sea level changes really exist. However, inter-annual or longer variability of sea level changes requires the corroboration and identification using more long-term data and more precisely scientific studies.

○ SEA LEVEL CHANGE MONITORING

Sea level changes are traditionally monitored using *in situ* measurements from along-shore tide gauges (Feng et al. 2013). Woodworth (1990) analyzed the long-term tide gauge records in European and then observed accelerations in individual gauges, although an overall slight deceleration of sea level changes from 1870 to 1990 is discovered. Douglas (1991) found that a rate of sea level rise is 1.8 mm/year, when the tide gauges longer than 50 years and far from tectonically active regions are selected for computation, which is larger than the rate of approximate 1 mm/year in pre-industrial era (prior to 1850), determined from the analysis of ocean sediments (Miller 2005). Nevertheless, many areas, in particular lakes located at remote regions, are rarely surveyed and even have no regular water gauge records. According to the last GPS measurements and studies, some tide gauges, for example in the Barents, Baltic and Caspian seas, have a positive vertical lift and may introduce remarkable errors in inter-annual variability of sea level if tide gauge records are used for sea level determination (Scherneck et al. 2001; Lebedev and Kostianoy 2008).

For these reasons, remote sensing techniques, such as satellite altimetry, appear as very complementary and useful tools (Mercier et al. 2002). The vertical motion of the earth's crust is not contained in inter-annual variations of the sea level changes since sea surface heights measured by satellite altimeters are relative to a reference ellipsoid (Kostianoy et al. 2011).

The era of precise altimetric sea level measurements began in 1992 with the launch of TOPEX/Poseidon (T/P) and in 2001 the Jason −1 (J-1) altimeter was launched to continue the T/P satellite mission (Leuliette et al. 2004). Combining data from these two altimeters allows scientists to investigate smaller features of ocean surfaces. The global sea surface heights are continuously recorded by the subsequent satellite missions, the Ocean Surface Topography Mission on the Jason-2 (OSTM/J-2) satellite launched in June 2008 and the ongoing Jason-3 mission.

Although satellite altimeters are primarily designed to study oceanic domains, their capability for monitoring continental water surfaces has already been demonstrated by many studies over a wide variety of water bodies; for example large lakes (Lebedev and Kostianoy 2008; Cretaux et al. 2011; Kropáceka et al. 2011; Kuo and Kao 2011; Imani et al. 2014a, b, c). Hwang et al. (2004) used T/P altimetric measurements to monitor inland lake level variations in China with good precision. Koblinsky et al. (1993) showed that the water level changes in Amazon Basin can be monitored by using U.S. Navy GEOSAT altimetric data. Morris and Gill (1994) also found that GEOSAT altimetric data could be used as a routine measurement for monitoring the temporal variation of water levels in the Great Lakes and St. Clair Lake. Alsdorf et al. (2001) used the spaceborne radar interferometric and T/P altimetric data to measure the water level changes and to calculate the volume of a big Amazon lake. Kuo at el. (2004) combined over 40-year long-term tide gauge records and 10-year T/P satellite altimetric measurements to estimate the vertical motions around the Baltic Sea. These studies show that satellite altimetry can provide a much faster, convenient, low costing, precise and new solution with a number of measurements to monitor inland waters.

◯ SEA LEVEL PREDICTIONS USING STATISTICAL APPROACHES ═══════

Monitoring and operational prediction of the oceans level variations are of great interest to many users. Although sea level monitoring is essentially useful for water management strategies, in most acceptable ecological and water related decisions, a modeling of sea level changes for prediction and analysis is usually necessary.

Many researchers have contributed their great efforts on the development and improvement of time series prediction models in the past several decades. An optimal forecasting technique is adoptable according to the availability of information and the time scale of the data. Although hydrologic models may provide some information, such as the availability and quality of water and impacts of human activities on land use change, to help decision-makers to deal with their concerning water problems, the reliability of prediction models relies on the model uncertainties, which normally are difficult to estimate (Yapo et al. 1996; Duan et al. 2003; Yang et al. 2007). Any attempt to predict climate change by atmospheric dynamical models requires a good understanding of how the water bodies operate and interact between factors like atmosphere, ocean, wind, topography, ice and snow melting etc. Since we do not really understand the whole climate change system, a modeling combining with the above mentioned factors may be difficult and not so suitable.

Dynamical models are unable to give an accurate simulation of the salient features of the mean seasonal monsoon and its inter-annual variability (Krishnamurti et al. 2000; Gadgil et al. 2005; Krishna Kumar et al. 2005; Wang et al. 2005). Rajeevan et al. (2004) showed that statistical models may perform better than the dynamical models in seasonal forecasts of southwest monsoon rainfall over India. Reed et al. (2008) have used several statistical approaches to determine the relative contributions of the short and long time scale variations using the long time series data of wind and water level from various sources and the result showed a good agreement between data and predictions using the autoregressive statistical models. Using statistical approaches, a complicated computational model can be replaced by a simpler approximation. In the field of oceanography, meteorology and hydrology, an easily operating statistical modeling of sea level variability is very helpful and forecasting can be done by taking the earlier values of sea level time series into consideration or making use of trigonometric functions to estimate the seasonal patterns of the data. Afterwards, the results of the models are evaluated by statistical criteria such as Root Mean Square Error (RMSE) of the differences between predictions and observations. The model with a minimum RMSE can be considered as the best one. Elsewhere, different statistical models for modeling/predictions of sea level changes have been successfully used and so statistical approaches are used to predict the water surface changes of Caspian Sea in this article.

⟲ CASE STUDY: CASPIAN SEA

Caspian Sea is located inside the Eurasian continent and its area is the greatest enclosed basin of the world. Its surface level is below the level of the World Ocean; at present, the increment is 27 m. The area of the sea covers approximately 390000 km^2 and the water volume reaches 78000 km^3 at a mean depth of 208 m (the maximum sea depth is about 1025 m). The sea extends over 1030 km from the north to the south and is 200 to 400 km wide from the east to the west (Fig. 1). The Volga River basin is the most important feeding of Caspian Sea and its area equals 1.4 million km^2 (Kostianoy and Kosarev 2005). The river runoff is the main contributor to the water level change of Caspian Sea.

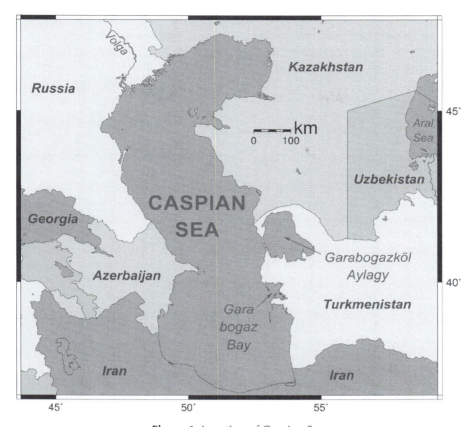

Figure 1 Location of Caspian Sea

Caspian Sea Level

The long-term sea level oscillations in Caspian Sea are especially important, since they are the principal influence on the basin and the socio-economic activities near its shores. Based on the historical, cartographic and paleogeographic studies, the changes of Caspian Sea area mainly result from water-climatic reasons, not from the tectonic movements (Kostianoy and Kosarev 2005). The range of the level changes over the past 2000 years reaches 7 m. The whole history of the water level changes of Caspian Sea can be found in Kostianoy and Kosarev (2005).

The level oscillations in Caspian Sea are a result of interrelated hydro-meteorological processes proceeding not only in its catchment area but also far beyond its bounding. Changes in the atmospheric circulation, in the Atlantic–European sector, in the hydro-meteorological conditions, in the Volga River basin, in the riverine runoff to Caspian Sea, the sea level variations of Caspian Sea and many other factors are the cause-and-effect relation. However, the relations between the circulation factors in the atmosphere and the meteorological regime in the Volga

River basin are the least studied. The formation of the Caspian Sea level is also strongly affected by the precipitation regime. Their interaction is very complicated. Therefore, the forecast possibility of the Caspian Sea level changes is basically a difficult problem, in particular while using a hydro-meteorological model by incorporating the factors mentioned above. If there is any error in the model, it may result in a huge mistake in prediction.

Prediction Approaches for Caspian Sea Level

Since Caspian Sea can be considered as a significant contribution to the climate change and water management scenarios in its adjacent countries and has a considerable effect on water-related decision making, a flexible and reliable prediction approach for its level changes is very much needed in order to reduce the effects of sea level changes and to avoid the influences of the incompleteness of the hydro-meteorological models.

Statistical methods, in particular the Artificial-Neural-Network approaches (ANNs), have been shown to be the flexible and reliable methods for the prediction of sea level changes by many studies. Vaziri (1997) investigated the monthly mean surface level fluctuations in the Caspian Sea using the period of January 1986 to December 1993 tide gauge records of Iranian Anzali Port by ANN and the autoregressive integrated moving average (ARIMA) methods. The results indicated that the ANN and ARIMA modeling are useful tools for the short-term predictions of time series data. Other researchers have also put effort into the prediction of the Caspian Sea level fluctuations by different methodologies in the last decades. Naidenov and Kozhevnikova (2000) applied a statistical method for the Caspian Sea level forecasting and Golitsyn (1995) predicted the Caspian Sea level using the balance method and found the most probable level from –27 m to –26 m in 2000. An adaptive technique has been employed by Ramezani Moziraji et al. (2010) for predicting the Caspian Sea level by combining a fuzzy concept with statistical logistic regression.

Recently, Imani et al. (2014 a, b, c) have presented the analysis and forecasting of Caspian Sea level pattern anomalies using T/P and J-1 altimetric data. Due to the large datasets and complicated analysis, in particular outlier detection from data, Imani et al. (2014a) used the principal component analysis (PCA) to reduce the complexity of analysis and applied ARIMA model for further analyzing and forecasting the sea level pattern anomalies in Caspian Sea. Afterwards, they introduced the time series obtained from the first principal component scores (sPC1) into the ARIMA modelling process.

An ARIMA model has the common form ARIMA (p, d, q) $(P, D, Q)s$, where (p, d, q) is the non-seasonal part and $(P, D, Q)s$ is the seasonal part of the model, as mentioned below.

$$\phi_p(B)\,\Phi_p(B^s)\nabla^d\nabla_S^D z_t = \theta_q(B)\Theta_Q(B^s)a_t, \tag{1}$$

where p is the order of non-seasonal autoregression; d is the number of regular differencing; q is the order of non-seasonal MA; P is the order of seasonal autoregression; D is the number of seasonal differencing; Q is the order of seasonal MA; B is the backward shift operator; s is the length of season (periodicity); ϕ is the AR operator of order p; Φ is the seasonal AR parameter of order P; ∇^d is the differencing operator; ∇_S^D is the seasonal differencing operator; z_t is the observed value at time point t; θ is the MA operator of order q; Θ is the seasonal MA parameter of order Q and a_t is the noise component of the stochastic model assumed to be $NID(0, \sigma^2)$[1] (Box et al. 1991 2007).

Based on time series derived from PCA analysis, they found that ARIMA (1, 1, 0) (0, 1, 1) model was the optimal fitting and prediction model in the Caspian Sea area for the T/P and J-1 altimetric data covering 1993-2008. As shown in Fig. 2, the sPC1 and fitted/predicted time series are very close to each other. It is clear that the pattern of sea level time series predicted from the model has the same seasonal fluctuations with the sPC1 series. In conclusion, their method could be thus a useful tool for the short-term predictions of the Caspian Sea level pattern anomalies.

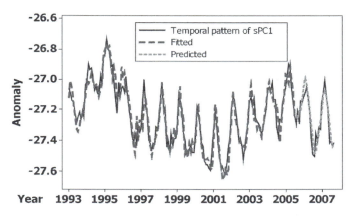

Figure 2 Line diagram showing the association between sPC1, fitted and predicted values from ARIMA (1, 1, 0) (0, 1, 1) (Imani et al. 2014a).

Despite the fact that conventional approaches such as ARIMA models showed a satisfactory fitting and prediction result in Caspian Sea and could be a useful tool for the short term predictions of sea level pattern anomalies, their time series was generated from an underlying linear process. However, these methods may not be always performed well when the hydrological time series are nonlinear. In order to provide more applicable statistical approach for nonlinear time series data, different artificial intelligent techniques were thus used for forecasting the short-term Caspian Sea level by the same authors (Imani et al. 2014b). ANNs can fit nonlinear problems very well and simulate believably the behaviour of complex systems without any preceding knowledge of the internal relations among their components. They tested and analyzed different ANN architectures and compared the outcomes with those derived from the conventional ARMA technique.

---

[1] $NID(0, \sigma^2)$ denotes normally and independently distributed with mean zero and variance σ^2.

Three multilayer structures of ANNs, namely, multilayer perceptron network (MLP), radial basis function (RBF) and generalized regression neural networks (GRNN), were developed for ANN processing here. The MLP is a feedforward ANN model that associates sets of input data with a set of appropriate output and its networks are an extension of perceptron networks containing one or more hidden layers (Cybenko 1989; Hornik et al. 1989). The RBF network is a feedforward neural network that consists of three layers, namely, the input, hidden and output layers. In RBF networks, the outputs are determined by computing the distance between the network inputs and the center of the hidden layer (Cichocki and Unbehauen 1993). Howlett and Jain (2001) gave a general expression of the RBF network representing the output–input relation as follows:

$$y_i = \sum_{j=1}^{M} \beta_j \phi_j(x) \qquad (2)$$

where y_i is the network output; M is the number of hidden neurons; x is the input data; β_j is the output layer weight of the RBF network and $\phi(x)$ is the Gaussian RBF given by

$$\phi_j(x) = \exp\left(-\frac{\|x - c_j\|^2}{\sigma_j^2} \right) \qquad (3)$$

where c_j is the centre and σ_j is the width of jth hidden neuron, respectively; $\|.\|$ denotes the Euclidean distance. The GRNN model is a kind of radial basis network that is often used for regression problems (Cichocki and Unbehauen 1993). The GRNN structure comprises four layers: the input, pattern, summation and output layers (Specht 1991).

The sea level anomalies from altimeters covering 1993-2008 were divided into two parts. The first part covering the period of 1993-2004 was used for the training procedure, whereas the second part covering 2005-2008 was utilized for the testing procedure. Imani et al. (2014b) tested different network structures for selecting the optimal model and found that the estimations by the RBF (5, 22, 1) network model produced the results with the minimum RMSE of 0.042 m and the highest R of 0.92 for the testing period (Table 1). This showed that the RBF (5, 22, 1) model was the best for forecasting results and agreed well with the observed sea level data (Fig. 3).

TABLE 1 Error statistics of the optimal models during the testing period (Imani et al. 2014b).

Type of model	MLP (5, 16-8, 1)	RBF (5, 22, 1)	GRNN (5, 0.3, 1)	ARMA (3, 3)
RMSE (m)	0.054	0.042	0.059	0.119
R	0.91	0.92	0.90	0.79

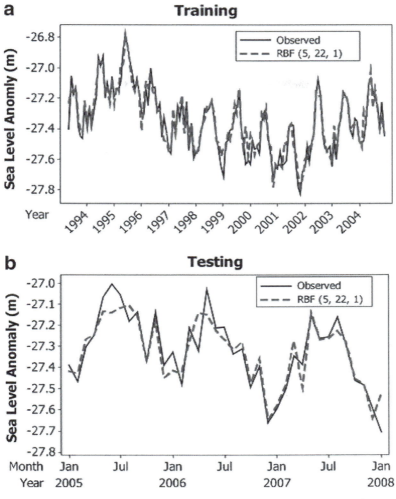

Figure 3 Observed and estimated sea level anomalies derived by the optimal RBF model during (a) the training and (b) testing periods (Imani et al. 2014b).

The RMSE of the ANN method was reduced by about 50% and R increased more than 15% compared with those values obtained from the ARMA method. This difference can be attributed to the fact that the ANN are flexible for the nonlinear time series data, while the ARMA model belongs to the class of linear models so it does not guarantee the prediction accuracy. Their results showed that sea level anomalies by ANN procedure can be predicted precisely and successfully without the use of any topographical details or other meteorological data as long as the required altimetric measurements are available.

Despite the numerous advantages of ANN architectures, the traditional ANNs have some defects as well. They generally suffer from considerable subjectivity in model architecture, its "black box" nature and larger computational time costs and maybe sometimes have over-fitting problems due to the experience of users. Recently, neural network methodologies, namely, SVMs have been applied for forecasting in time series analysis (e.g. Vapnik et al. 1997; Cristianini and Taylor 2000; Kim 2003; Guven and Gunal 2008; Rajasekaran et al. 2008; Ghorbani et al. 2010; Lins et al. 2013). SVMs is an advanced neural network technology based on statistical learning. It is one kind of the computational intelligence approach which is an offshoot of artificial intelligence (AI) composed of adaptive mechanisms and learning ability that facilitate intelligent behavior in a complex environment (Islam et al. 2014).

The learning algorithms of SVMs, developed by Vapnik et al. (1997), are described specifically by the capacity control of the decision function, the kernel functions and the sparsity of the solution (Cristianini and Taylor 2000). SVMs implement the structural risk minimization principle instead of the empirical risk minimization principle in the traditional neural network models. The structural risk minimization principle is able to minimize the upper bound of the generalization error rather than minimize the training error (Tay and Cao 2001). SVMs are very resistant to the over-fitting problem so that it has highly generalized performance in solving various time series forecasting cases. Besides, unlike the most cases of AI approaches, such as ANN algorithms, SVMs does not need a large number of training examples for obtaining a better optimization (Islam et al. 2012).

The support vector regression (SVR) is such a neural network methodology and is developed to deal with nonlinear problems. The SVR formalism considers the following regression function (Vapnik 1999, 2000):

$$f(x) = (v \cdot \Phi(x)) + b, \qquad (4)$$

where v is the weight vector; b is a constant; $\Phi(x)$ denotes a mapping function in the feature space and $(v \cdot \Phi(x))$ describes the dot production in the feature space f. In fact, the problem of nonlinear regression in the lower-dimensional input space (x) is transformed into a linear regression problem in a high-dimensional feature space.

Despite well-documented studies in other fields, the applications of SVMs in hydrology are still few. Sivapragasam et al. (2001) conducted one-lead-day rainfall and runoff forecasting using SVMs and Tripathi et al. (2006) applied SVMs in the statistical downscaling of precipitation at a monthly timescale. In 2013, Srivastava et al. (2013a, b) investigated different AI techniques including ANNs, SVMs, relevance vector machines and generalized linear models to downscale soil Moisture and Ocean Salinity soil moisture content and later evaluated four data fusion techniques, such as linear weighted algorithm, multiple linear regression, Kalman filter and ANN for Soil Moisture Deficit estimation using SMOS satellite and WRF-NOAH land surface model derived soil moisture.

To the best of our knowledge, the recent available study on SVMs application in sea level time series prediction has been performed by Imani et al. (2014c) following their previous study regarding the traditional ANN architectures. They applied the SVMs approaches to predict the Caspian Sea level anomalies derived from the T/P, J-1 and J-2 satellite missions covering the period of June 1992 to December 2013. Fig. 4 shows that the SVR prediction results obtained by Imani et al. (2014c) had very well forecast performance and highlighted the capability of the SVR to predict the seasonal changes and to determine the general behaviour of the time series. Their analysis also showed the high correlation between the observed and predicted data (the coefficient of determination $R^2 = 0.99$).

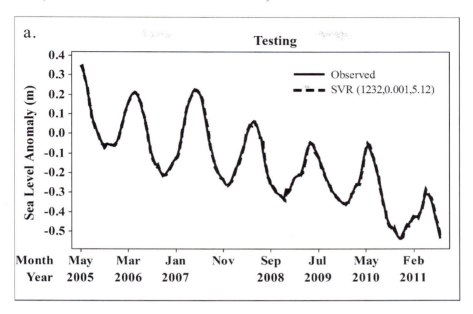

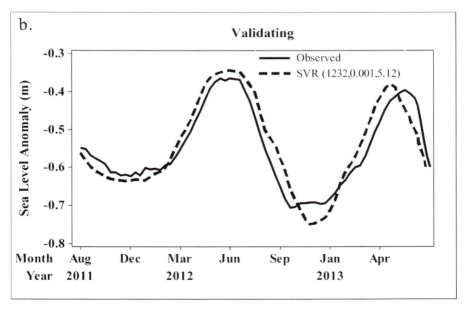

Figure 4 Observed and estimated sea level anomalies of SVR during the (a) testing and (b) predicting periods (Imani et al. 2014c).

○ CONCLUSIONS

In this article, we have given the recent studies in the discussion and prediction of the short-term Caspian Sea level variations using the T/P, J-1 and J-2/OSTM altimetric data. Based on our analysis, SVMs has the best performance in predicting the short-term Caspian Sea level anomalies compared to the ARIMA models and the traditional ANNs. For the study of the inter-annual or longer variability of sea level changes, more long-term data including data of decades and even several centurial data is naturally required.

Overall, the effort has been put in this article to present the promising forecasting approaches of the most commonly available models from traditional to the state-of-the-art methods in solving site-specific and real-time water resource problems, with the introduction of the SVMs as an applicable alternative for sea level time series forecasting. The provided information in this article may thus lead to a better understanding of applicable tools in forecasting stochastic time series and an effective insight for more precise prediction-based decision making in water management scenarios.

○ REFERENCES

Alsdorf, D., Birkett, C. and Dunne, T. 2001. Water level changes in a large Amazon lake measured with spaceborne radar interferometry and altimetry. Geophysical Research Letter 28: 2671-2674.

Bindoff, N.L. and Willebrand, J. 2007. Chapter 5: Observations: oceanic climate change and sea level. Section 5.5.2.4: Inter-annual and decadal variability and long-term changes in sea level. In IPCC Fourth Assessment Report: Climate Change 2007: Working Group I: Physical Science Basis, Cambridge University Press, Cambridge.

Box, G.E.P., Jenkins, G.M. and Reinsel, G.C. 1991. Time series analysis, forecasting and control. Prentice Hall, Englewood Cliffs, New Jersey.

Box, G.E.P., Jenkins, G.M. and Reinsel, G.C. 2007. Time series analysis: forecasting and control, 3rd Edition. Dorling Kindersley (India) Pvt Ltd, New Delhi, India (licensees of Pearson Education in South Asia).

Brunel, C. and Sabatier, F. 2009. Potential influence of sea-level rise in controlling shoreline position on the French Mediterranean Coast. Geomorphology 107(1-2): 47-57.

Church, J., Woodworth, A., Aarup, T. and Wilson, W.S. (eds.). 2010. Understanding Sea-Level Rise and Variability. Wiley-Blackwell, Oxford.

Cichocki, A. and Unbehauen, R. 1993. Neural Networks for Optimization and Signal Processing. Wiley, Chichester.

Cretaux, J.F., Jelinski, W., Calmant, S., Kouraev, V. and Vuglinski, M.V. 2011. SOLS: a lake database to monitor in the near real time water level and storage variations from remote sensing data. Advances in Space Research 47: 1497-1507.

Cristianini, N. and Taylor, J.S. 2000. An Introduction to Support Vector Machines and Other Kernel-based Learning Methods. Cambridge University Press, New York.

Cybenko, G. 1989. Approximation by superposition of a sigmoidal function. Mathematics of Control, Signals and Systems 2(4): 303-314.

Douglas, B.C. 1991. Global sea level rise. Journal of Geophysical Research 96: 6981-6992.

Duan, Q., Gupta, H.V., Sorooshian, S., Rousseau, A.N. and Turcotte, R. (eds.). 2003. Calibration of Watershed Models. American Geophysical Union, Washington, D.C.

Feng, G., Jin, S. and Zhang, T. 2013. Coastal sea level changes in Europe from GPS, tide gauge, satellite altimetry and GRACE, 1993-2011. Advances in Space Research 51(6): 1019-1028.

Gadgil, S., Rajeevan, M. and Nanjundiah, R. 2005. Monsoon prediction – why yet another failure. Current Science 88(9): 1389-1400.

Ghorbani, M.A., Khatibi, R., Aytek, A., Makarynskyy, O. and Shiri, J. 2010. Sea water level forecasting using genetic programming and comparing the performance with artificial neural networks. Computers & Geosciences 36: 620-627.

Golitsyn, G.S. 1995. The Caspian Sea level as a problem of diagnosis and prognosis of the regional climate change. Atmospheric and Oceanic Physics 31: 366-372.

Guven, A. and Gunal, M. 2008. Genetic programming approach for prediction of local scour downstream hydraulic structures. Journal of Irrigation and Drainage Engineering 134(2): 241-249.

Hornik, K., Stinchcombe, M. and White, H. 1989. Multilayer feed forward networks are universal approximators. Neural Networks 2(5): 359-366.

Howlett, R.J. and Jain, L.C. 2001. Radial Basis Function Networks 1: Recent Developments in Theory and Applications. Physica-Verlag, New York.

Hwang, C.W., Peng, M.F., Ning, J.S. and Sui, C.H. 2004. Lake level variations in China from TOPEX/Poseidon altimetry. DOI: 10.1111/j.1365-246X.2005.02518.x.

Imani, M., You, R.J. and Kuo, C.Y. 2014a. Analysis and prediction of Caspian Sea level pattern anomalies observed by satellite altimetry using autoregressive integrated moving average models. Arabian Journal of Geosciences 7: 3339-3348.

Imani, M., You, R.J. and Kuo, C.Y. 2014b. Caspian Sea level prediction using artificial neural networks (ANNs) and satellite altimetry. International Journal of Environmental Science and Technology 11(4): 1035-1042.

Imani, M., You, R.J. and Kuo, C.Y. 2014c. Forecasting Caspian Sea level changes using satellite altimetry data (June 1992 – December 2013) based on evolutionary support vector regression algorithms and gene expression programming. Global and Planetary Change 121: 53-63.

Islam, T., Rico-Ramirez, M.A., Han, D. and Srivastava, P.K. 2012. Artificial intelligence techniques for Klutter identification with polarimetric radar signatures. Atmospheric Research 109: 95-113.

Islam, T., Srivastava, P.K., Gupta, M., Zhu, X. and Mukherjee, S. (eds.). 2014. Computational Intelligence Techniques in Earth and Environmental Sciences. Springer Verlag, Dordrecht.

Jevrejeva, S., Moore, J.C., Grinsted, A. and Woodworth, P.L. 2008. Recent global sea level acceleration started over 200 years ago. Geophysical Research Letters 35, L08715, DOI: 10,1029,2008 GL033611.

Kim, K.J. 2003. Financial time series forecasting using support vector machines. Neurocomputing 55: 307-319.

Koblinsky, C.J., Clarke, R.T., Brenner, A.C. and Frey, H. 1993. Measurement of river level variations with satellite altimetry. Water Resource Research 29(6): 1839-1848.

Kostianoy, A.G. and Kosarev, A.N. (eds.). 2005. The Caspian Sea Environment. Springer Verlag, Berlin.

Kostianoy, A.G., Lebedev, S.A. and Solovyov, D.M. 2011. Satellite monitoring of water resources in Turkmenistan. The Fifteenth International Water Technology Conference, IWTC-15, 2011, Alexandria, Egypt.

Krishna Kumar, K., Hoerling, M. and Rajagopalan, B. 2005. Advancing dynamical prediction of Indian Monsoon Rainfall. Geophysical Research Letter 32(8): 1-4, L08704, 10.1029/2004GL021979, 4.

Krishnamurti, T.N., Kishtawal, C.M., LaRow, T.E., Bachiochi, D.R., Zhang, Z., Williford, C.E., Gadgil, S. and Surendran, S. 2000. Multimodel ensemble forecasts for weather and seasonal climate. Journal of Climate 13: 4196-4216.

Kropáceka, J., Brauna, A., Kang, C., Feng, C., Yeb, Q. and Hochschild, V. 2011. Analysis of lake level changes in Nam Co in central Tibet utilizing synergistic satellite altimetry and optical imagery. International Journal of Applied Earth Observation and Geoinformation. DOI: 10.1016/ j.jag.2011.10.001.

Kuo, C.Y. and Kao, H.C. 2011. Retracked Jason-2 altimetry over small water bodies: case study of Bajhang river, Taiwan. Marine Geodesy 34(3-4): 382-392.

Kuo, C.Y., Shum, C.K., Braun, A. and Mitrovica, J.X. 2004. Vertical crustal motion determined by satellite altimetry and tide gauge data in Fennoscandia. Geophysical Research Letters 31, l01608, DOI: 10.1029/2003gl019106.

Lebedev, S.A. and Kostianoy, A.G. 2008. Integrated use of satellite altimetry in the investigation of the meteorological, hydrological and hydrodynamic regime of the Caspian Sea. Terrestrial Atmospheric and Oceanic Science 19(1-2): 71-82.

Leuliette, E.W., Nerem, R.S. and Mitchum, G.T. 2004. Results of TOPEX/Poseidon and Jason-1 calibration to construct a continuous record of mean sea level. Marine Geodesy 27: 79-94.

Lins, I.D., Araujo, M., Moura, M.d.C., Silva, M.A. and Droguett, E.L. 2013. Prediction of sea surface temperature in the tropical Atlantic by support vector machines. Computational Statistics & Data Analysis, 61: 187-198.

Mercier, F., Cazenave, A. and Maheu, C. 2002. Interannual lake level fluctuations in Africa (1993–1999) from TOPEX-Poseidon: connections with ocean-atmosphere interactions over the Indian ocean. Global and Planetary Change 32: 141-163. DOI: 10.1016/ S0921-8181(01)00139-4.

Miller, K.G. 2005. The phanerozoic record of global sea-level change. Science 310: 1293-1298.

Milne, G.A., Gehrels, W.R., Hughes, C.W. and Tamisiea, M.E. 2009. Identifying the causes of sea-level change. Nature Geoscience 2: 471-478.

Morris, C.S. and Gill, S.K. 1994. Variation of Great-lakes water levels derived from GEOSAT altimetry. Water Resource Research 30(4): 1009-1017.

Naidenov, V.I. and Kozhevnikova, I.A. 2000. Nonlinear variations of the level of the Caspian Sea and the global climate. Doklady Physics 4(5): 340-345.

Pycroft, J., Abrel, J. and Ciscar, J.C. 2015. The global impacts of extreme sea-level rise: a comprehensive economic assessment. Environmental and Resource Economics, DOI 10.1007/s10640-014-9866-9.

Rajasekaran, S., Gayathri, S. and Lee, T.L. 2008. Support vector regression methodology for storm surge predictions. Ocean Engineering 35: 1578-1587.

Rajeevan, M., Pai, D.S., Dikshit, S.K. and Kelker, R.R. 2004. IMD's new operational models for long range forecast of south-west Monsoon Rainfall over India and their verification for 2003. Current Science 86(3): 422-431.

Ramezani Moziraji, F., Yaghobi, M. and Zargari Kordkolaii, J. 2010. Caspian Sea level prediction by auto fuzzy regression. World Applied Sciences Journal 8(3): 288-292.

Reed, R.E., Dickey, D., Burkholder, J.M., Kinde, C.A. and Brownie, C. 2008. Water level variations in the Neuse and Pamlico estuaries, north Carolina due to local and remote forcing. Estuarine, Coastal and Shelf Science 76: 431-446.

Sachs, J.D., Mellinger, A.D. and Gallup, J.L. 2001. The geography of poverty and wealth. Scientific America 284(3): 70-75.

Scherneck, H.G., Johansson, J.M., Vermeer, M., Davis, J.L., Milne, G.A. and Mitrovica, J.X. 2001. BIFROST project: 3-D crustal deformation rates derived from GPS confirm postglacial rebound in Fennoscandia. Earth Planets and Space 53: 703-708.

Senapati, S. and Gupta, V. 2014. Climate change and coastal ecosystem in India: issues in perspectives. International Journal of Environmental Sciences 5(3): 530-543.

Singh, M.K. and Singh, B.R. 2013. Modeling of future sea level rise through melting glaciers. All India Seminar on Sources of Planet Energy, Environmental & Disaster Science: Challenges and Strategies (SPEEDS-2013) on September 07-08, 2013, Lucknow.

Sivapragasam, C., Liong, S.Y. and Pasha, M.F.K. 2001. Rainfall and runoff forecasting with SSA–SVM approach. J. Hydroinformatics 3(7): 141-152.

Small, C. and Nicholls, R.J. 2003. A global analysis of human settlement in coastal zones, Journal of Coastal Research 19(3): 584-599.

Specht, D.F. 1991. Enhancements to probabilistic neural network. In Proceedings of the International Joint Conference Neural Network 1: 761-768.

Srivastava, P.K., Han, D., Ramirez, M.R. and Islam, T. 2013a. Machine learning techniques for downscaling SMOS satellite soil moisture using MODIS land surface temperature for hydrological application. Water Resources Management 27(8): 3127-3144.

Srivastava, P.K., Han, D., Rico-Ramirez, M.A., Al-Shrafany, D. and Islam, T. 2013b. Data fusion techniques for improving soil moisture deficit using SMOS satellite and WRF-NOAH land surface model. Water Resources Management 27(15): 5069-5087.

Svetlana, J., Moore, J.C., Grinsted, A. and Woodworth, P.L. 2008. Recent global sea level acceleration started over 200 years ago? Geophysical Research Letters 35(8), DOI: 10.1029/2008 GL033611.

Takagi, H., Ty, T.V., Thao, N.D. and Esteban, M. 2014. Ocean tides and the influence of sea-level rise on floods in urban areas of the Mekong delta. Journal of Flood Risk Management, DOI: 10.1111/jfr3.12094.

Tay, F.E.H. and Cao, L.J. 2001. A comparative study of saliency analysis and genetic algorithm for feature selection in support vector machines. Intelligent Data Analysis 5(3): 191-209.

Tripathi, S., Srinivas, V.V. and Nanjundiah, R.S. 2006. Downscaling of precipitation for climate change ccenarios: a support vector machine approach. Journal of Hydrology 330: 621-640.

Turner, R.K., Subak, S. and Adger, N.W. 1996. Pressures, trends and impacts in coastal zones: Interactions between Socioeconomic and Natural Systems. Environmental Management 20: 159-173.

Vapnik, V., Golowich, S. and Smola, A. 1997. Support vector method for function approximation, regression estimation and signal processing. In: Mozer, M., Jordan, M. and Petsche, T. (eds.). Advances in Neural Information Processing Systems. MIT Press, Cambridge, Massachusetts, 9: 281-287.

Vapnik, V.N. 1999. An overview of statistical learning theory. IEEE Transactions on Neural Networks 10(5): 988-999.

Vapnik, V.N. 2000. The Nature of Statistical Learning Theory. Springer Verlag, New York.

Vaziri, M. 1997. Predicting Caspian Sea surface water level by ANN and ARIMA models. Journal of Waterway, Port, Coastal and Ocean Engineering 123: 158-162.

Wang, Y., Sun, G. and Liao, M. 2005. Using MODIS images to examine the surface extents and variations derived from the DEM and laser altimeter data in the Danjiangkou Reservoir, China. International Journal of Remote Sensing 29(1): 293-311.

Woodworth, P.L. 1990. A search for accelerations in records of European mean sea level. International Journal of Climatology 10: 129-143.

Yang, J., Reichert, P., Abbaspour, K.C. and Yang, H. 2007. Hydrological modelling of the Chaohe Basin in China: statistical model formulation and Bayesian inference. Journal of Hydrology 340(304): 167-182.

Yapo, P.O., Gupta, H.V. and Sorooshian, S. 1996. Automatic calibration of conceptual rainfall runoff models: sensitivity to calibration data. Journal of Hydrology 181(1-4): 23-48.

16
CHAPTER

◇◇◇

Spatio-temporal Uncertainty Model for Radar Rainfall

Qiang Dai,[1,2,3,*] *Dawei Han,*[2] *Miguel A. Rico-Ramirez*[2] and *Prashant K. Srivastava*[4,5,6]

ABSTRACT

Weather radar has been widely used in hydrologic forecasting and decision making; nevertheless, there is increasing attention on its uncertainties that propagates through hydrologic models. This chapter proposes a fully formulated uncertainty model that can statistically quantify the characteristics of the radar rainfall errors and their spatial and temporal structure, which is a novel method of its kind in the radar data uncertainty field. The uncertainty model is established based on the distribution of gauge rainfall conditioned on radar rainfall (GR|RR). Its spatial and temporal dependences are simulated based on the copula function. With this proposed uncertainty model, a Multivariate Distributed Ensemble Generator (MDEG) driven by the copula and autoregressive filter is designed. As wind is a typical weather factor that influences radar measurement, this study introduces the wind field into the uncertainty model and designs the radar rainfall uncertainty model under different wind conditions. The Brue catchment (135 sq. km) in Southwest England covering 28 radar pixels and 49 rain gauges is chosen as the experimental domain for this study.

KEYWORDS: uncertainty model, radar rainfall, MDEG, wind conditions, copula, autoregressive filter.

[1] Key Laboratory of VGE of Ministry of Education, Nanjing Normal University, Nanjing, China.
[2] Water and Environmental Management Research Centre, Department of Civil Engineering, University of Bristol, Bristol, UK.
[3] Jiangsu Center for Collaborative Innovation in Geographical Information Resource Development and Application, Nanjing, China.
[4] Hydrological Sciences, NASA Goddard Space Flight Center, Greenbelt, Maryland, USA.
[5] Earth System Science Interdisciplinary Center, University of Maryland, Maryland, USA.
[6] Institute of Environment and Sustainable Development, Banaras Hindu University, Varanasi, India.
* Corresponding author: q.dai@bristol.ac.uk

○ INTRODUCTION

Weather radar, with its advantage of providing observation of the precipitation with high spatial and temporal resolutions and good areal coverage, has gradually been accepted as an important input in hydrological applications during the past half century (He et al. 2011; Collier 1986' Wood et al. 2000). However, as it measures rainfall remotely and indirectly, there is large uncertainty associated with radar rainfall measurements (Villarini and Krajewski 2010; Kitchen et al. 1994; Harrold et al. 1974). Enormous studies have been presented to solve these potential uncertainties since the appearance of the weather radar (Villarini and Krajewski 2010). The quality of radar rainfall product has seen huge improvement and it can be used in most hydrological applications with relatively high confidence. However, compared to ground observations such as rain gauge, its quality is still very poor. For this reason, the uncertainty model of radar rainfall is established with the assistance of gauge measurements, no matter if it is a simple linear model or a complicated error model like the proposed scheme in this work (Dai et al. 2015). This is almost an essential step in dealing with radar rainfall products. These radar rainfall uncertainty models can be classified into two types, namely the empirical uncertainty model and the real-time uncertainty correction. The recent studies have focused on the empirical uncertainty model and many statistical models have been proposed (Ciach et al. 2007; Germann et al. 2009). However, some researchers may debate the predictability of radar uncertainty models and argue that the characteristics of the measurement uncertainty are varied in each individual storm. The most serious concern associated with the empirical model is that different synoptic regimes may induce different behaviors of radar measured uncertainties (Dai et al. 2015). So the empirical error model should be established with consideration of specific regimes. Based on these evidences, this study presents the first fully formulated radar rainfall uncertainty model based on conditional distribution, copula function and autoregressive model with long-term radar-gauge datasets and studies its characteristics under different synoptic regimes. The influence of wind is mainly presented herein. Other affected factors such as seasons and storm types are also discussed.

DATA AND METHODS

Study Area and Data Sources

A catchment over temperate maritime climate located in Somerset, south-west England (51.08°N and 2.58°W), is chosen as the experimental catchment for this study. It covers an area of 135 sq. km to its river gauging station at Lovington (Moore et al. 2000; Roberts et al. 2000) and its elevation is modest, from 35 m to 190 m above the sea level (Wheater et al. 1999). The map of the catchment and 49 rain gauges network grids are plotted in Fig. 1. The radar data are from the Wardon Hill radar with 2 km Cartesian grids, located at a range of around 40 km from the centre of the catchment. It can be observed from Fig. 1 that the Brue catchment overlaps with radar grids by 52 squares and 28 pixels of them are covered with the most area. The gauge rainfall is from a dense network of 49 standard Cassella tipping bucket gauges (TBRs) with 0.2 mm resolution and time resolution of 10 seconds (Wood et al. 2000). An automatic weather station (AWS) is located in the lowland sub-network of the catchment. It records wind speed and direction, solar and net radiation, wet and dry bulb temperature, atmospheric pressure and rainfall (0.2 mm tip) every 15 minutes. Among all these variables, the wind speed is mainly used in this study. The hourly accumulated radar rainfall, gauge rainfall and wind data are available from October 1993 to February 2000, which are provided by the Hydrology Radar Experiment (HYREX) conducted by the Natural Environment Research Council (NERC) Special Topic Programme.

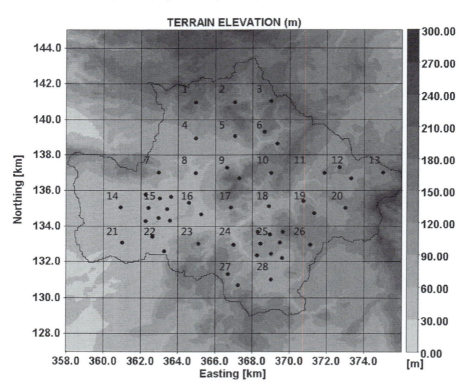

Figure 1 Map of the rain gauge network and the Brue catchment. The blue dots represent the rain gauge and the black grid represents the radar pixels. The number in the pixel refers to the index of the corresponding pixel (Dai et al. 2014a).

Radar-rainfall Uncertainty Model

Generally, the major goals of the uncertainty model include: 1) calculation of the probability distribution of 'true' rainfall once given a radar estimate, which reveals the degree of confidence in the radar products; 2) investigation of the relationship between radar-rainfall uncertainty and rainfall intensity; 3) analysis of the characteristics of the spatial and temporal dependence of the uncertainty model; 4) investigation of other possible factors that affect radar-rainfall uncertainty such as wind, seasons and storm types. In this study, we generate a great number of random fields to represent the radar-rainfall uncertainty based on the Monte Carlo method. We assume the true areal averaged pixel-scale rainfall is composed of two components, namely the deterministic distortion component and the random component (Dai et al. 2014b). Thus the strictly statistical representation of the model is expressed as:

$$\psi_{t,i} = h_t(R_t) + \varepsilon_{t,i}(R_t) \qquad (1)$$

where $\psi_{t,i}$ is a realistic "true rainfall" with respect to the radar measurement R_t for member i time step t. $h_t(R_t)$ and $\varepsilon_{t,i}(R_t)$ represent the deterministic component and random component of "true rainfall" respectively. We firstly generate a series of possible random components and then add them into the deterministic component. So the challenge of generating ensembles lies in the simulation of the distribution of random errors. As the random error is the difference between the deterministic component and the true rainfall, the distribution of random errors can be obtained from the distribution of Gauge Rainfall (GR) conditioned on the Radar Rainfall (RR) estimates. A schematic diagram of the process to generate GR|RR distribution is shown in Fig. 2.

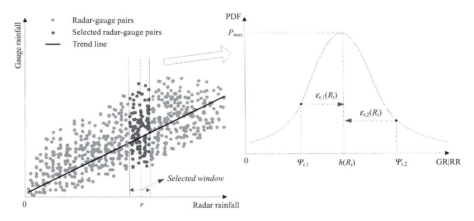

Figure 2 Schematic diagram of scatter points of radar-gauge rainfall pairs and probability density function (PDF) of GR|RR according to the empirical distribution. In the scatter points figure, r refers to a radar rainfall estimate (Dai et al. 2015).

To better apply the uncertainty model in the proposed ensemble generator, the variance of random errors and the deterministic component of radar rainfall are parameterized using power law functions:

$$\sigma_\varepsilon(x, y) = a_\varepsilon R(x, y)^{b_\varepsilon} \qquad (2)$$

$$h(x, y) = a_h R(x, y)^{b_h} \qquad (3)$$

where the $R(x, y)$ represents the rainfall variable for the point location at (x, y). The a_h, b_h, a_ε and b_ε are the coefficients of the parameterized model for the deterministic component and the variance of the random error. They depend crucially on the spatial location.

Once we have modeled the marginal distribution of radar rainfall uncertainty, we should ensure the generated ensemble rainfalls maintain the spatial and temporal correlation over the whole catchment. A new scheme called Multivariate Distributed Ensemble Generator (MDEG) is designed, driven by the copula and autoregressive filter to generate the desired ensemble rainfall. The main idea of MDEG is to generate a number of random errors of rainfall with modeled marginal distribution and spatial dependence by the copula and then impose the random error with a desired temporal dependence using the autoregressive filter.

The set of parameters of copula are obtained through a system of estimating formulas, with each formula being a score function from the marginal distribution (Joe 1997; Dai et al. 2014b). The parameters of copula function can be estimated using the Inference Function for Margins (IFM) method (Xu 1996; Godambe 1960; Godambe 1976 Joe 1997). With copula parameters, a vector of uniform random variable $u_t(u_{t, 1}, \ldots u_t, n)$ with marginal distributions of U $(U_1, \ldots U_n)$ $(U_i \in [0, 1])$ is generated using the copula function. The random vector is obtained from the standard multivariate distribution or conditional distribution. We have compared four copulas from elliptical copulas and Archimedean copulas (Gaussian, t copulas, Clayton and Gumbel copulas) and found t-copula has the best performance in estimating the spatio-temporal dependence of the random error (Dai et al. 2014b). The density function of the t-copula with v degrees of freedom and ρ correlation matrix is expressed as (Malevergne and Sornette 2003):

$$c(u_1, u_n) = \frac{1}{\sqrt{\det \rho}} \frac{\Gamma((v + n)/2)\Gamma(v/2)^{n-1}}{(\Gamma(v + 1)/2)^n} \frac{\prod_{k=1}^{n}(1 + y_k^2/v)^{(v+1)/2}}{(1 + y'\rho^{-1}y/v)^{(v+n)/2}} \quad (4)$$

where t_v represents the univariate t-distribution and y_k is the inverse of the t_v with the marginal cumulative distribution function (CDF) u_k. Having these multivariate uniform random fields, we can obtain the random error with the desirable marginal distributions by the probability integral transform. Then the autoregressive filter is used to impose the desired temporal correlation. A second-order autoregressive model AR (2) (Priestley 1981) is selected in this study since it is causal and easy to estimate in real time (Seed 2003). In the final step, the random error that satisfies the spatial and temporal structure is superimposed onto the original deterministic component using Equation (1). This procedure is repeated until the end of the event.

Impact of Synoptic Regimes on Radar-rainfall Uncertainty

As stated earlier, wind has huge impact on radar rainfall uncertainty, so we study the dependence of radar rainfall uncertainty on wind field. We can regard wind speed as a variable in Equation (1) to incorporate the wind field to the MDEG model. In such a situation, we need to construct the distribution of gauge rainfall conditioned on both the wind speed (WS) and radar rainfall (GR|RR∩WS). However, it is difficult to implement it as the size of subsample that satisfies a given wind speed and rainfall rate would be quite small (Dai et al. 2015). Instead, we can assume only that the magnitude of wind speed induces considerable changes of behavior in the uncertainty model. Thus the original datasets are classified by three subsamples, namely WDI (0-2 m/s), WDII (2-4 m/s) and WDIII (>4 m/s). The ensemble generator is established and implemented under different subsamples and configured to run the model based on the current wind speed in a real-time situation.

To illustrate the variation of rainfall uncertainties under different wind conditions, the spread of the bands and ensemble bias are calculated and compared, which are defined as:

$$\delta_D = \frac{1}{nt} \sum_{i=1}^{nt} [P_{95}(t) - P_5(t)] \quad P_{50}(t) > 0 \tag{5}$$

$$\delta_B = \sum_{i=1}^{n} |P_{50}(t) - P^{ref}(t)| \tag{6}$$

where nt is the number of time steps corresponding to P_{50} larger than zero, while $P_{95}(t)$, $P_5(t)$ and $P_{50}(t)$ are the 95th, 5th and 50th percentiles of the ensemble members at time t respectively. n is the number of time steps.

Except wind field, other factors such as seasons and storm type are also worth remarking. The uncertainty model established for each season is easily implemented with the similar method as wind factor. For storm type, we believe they also have significant effect on radar measurement. There are two possible methods of classifying storm type: spatial and temporal evenness and lamb weather type. Surely, there are many other factors worth integrating into the error model. One may argue that it is impossible to consider all synoptic factors, especially for some complicated elements such as pressure and relative humidity of air. However, not all synoptic factors have critical impact on rainfall measurement (Dai et al. 2015). Wind field, seasons and storm types are mainly acknowledged in the published literatures.

○ RESULTS AND DISCUSSION ════════════════════════

Generation of Ensemble Rainfalls

Error Model Implementation

In this case study, the radar and gauge data from 1993 to 1997 is used to establish the uncertainty model, while the 30 typical events from the remaining dataset (1998-2000) are chosen to generate ensemble rainfall estimates. We take into account rainfall accumulation and wind speed when we select the events. The detailed information (includes durations, accumulated rainfall and mean wind speed) of these events is listed in Table 1.

TABLE 1 Basic information of 30 storm events.

Event ID	Start date	End date	Accumulated rainfall (mm)	Mean wind speed (m/s)
WD I-1	1998-09-10:21	1998-09-11:07	24.03	0.46
WD I-2	1999-08-03:01	1999-08-03:07	24.09	0.46
WD I-3	1999-08-18:04	1999-08-19:05	59.04	1.13
WD I-4	2000-01-03:08	2000-01-03:21	45.82	1.43
WD I-5	1998-12-15:12	1998-12-15:19	25.33	1.36
WD I-6	1998-12-23:05	1998-12-23:18	68.00	1.02
WD I-7	1999-02-08:05	1999-02-08:17	58.29	1.42
WD I-8	1999-04-23:00	1999-04-24:15	133.02	1.89
WD I-9	1999-08-09:07	1999-08-10:03	83.59	1.34
WD I-10	1999-09-22:07	1999-09-22:14	18.84	1.73
WD II-1	1998-01-18:01	1998-01-18:18	104.20	3.25
WD II-2	1998-10-16:18	1998-10-17:08	114.10	2.30
WD II-3	1999-12-22:11	1999-12-23:00	132.86	2.60
WD II-4	2000-01-05:18	2000-01-06:05	43.38	2.87
WD II-5	1998-10-20:22	1998-10-21:06	37.44	3.19
WD II-6	1998-10-22:21	1998-10-23:06	54.66	3.48
WD II-7	1998-11-11:20	1998-11-12:14	75.88	2.63
WD II-8	1999-01-19:03	1999-01-20:07	86.32	3.43
WD II-9	1999-12-17:03	1999-12-17:10	68.63	3.63
WD II-10	2000-02-07:07	2000-02-07:15	76.74	3.73
WD III-1	1998-03-03:03	1998-03-03:10	45.63	4.17
WD III-2	1999-12-23:20	1999-12-24:05	102.18	4.00
WD III-3	1999-12-24:12	1999-12-24:19	84.34	4.13
WD III-4	2000-02-10:04	2000-02-10:11	82.75	4.51
WD III-5	1998-04-04:10	1998-04-04:16	27.16	4.02
WD III-6	1998-10-25:06	1998-10-25:12	54.46	4.09
WD III-7	1998-12-26:09	1998-12-26:21	52.55	4.18
WD III-8	1999-01-15:10	1999-01-16:10	105.68	4.15
WD III-9	1999-01-26:13	1999-01-26:23	67.61	4.61
WD III-10	2000-02-12:05	2000-02-12:12	49.30	4.51

The implementation of the MDEG is divided into two main parts: namely establishing an uncertainty model and running the uncertainty model in real time. The core part of the error model is to estimate the marginal distribution, spatial correlation among different lag distances and temporal correlation among different lag times. The parameters of deterministic component and the variance of the random component are listed in Table 2.

TABLE 2 Parameters of the deterministic component and variance of random error.

Wind domain	Deterministic		Variance of random	
	a_h	b_h	a_ε	b_ε
WD I	1.27	0.73	0.68	0.91
WD II	1.25	0.67	0.76	0.70
WD III	1.15	0.63	0.86	0.23
Entire	1.25	0.69	0.73	0.77

The spatial and temporal correlations under different wind ranges are plotted in Figs. 3 and 4. The degree of freedom of t-copula functions are listed in Table 3. To help visual comparisons among correlations, the points in the scatter plots of spatial correlations are fixed into a three-parameter exponential function. The estimated parameters are drawn in Fig. 3.

TABLE 3 Estimated degree of freedom of t-copula functions. The spatial and temporal dependence for the three wind ranges and the entire dataset are estimated by the t-copula separately.

	WD I	WD II	WD III	Entire
Spatial correlation	1.64	1.68	1.99	1.63
Temporal correlation	2.96	3.84	4.18	2.80

In terms of the temporal correlation, it is observed from Fig. 4 that there are small differences in correlation coefficients among the four samples. The deviations become larger with the increase of lag time. There is no wind range that always has the highest temporal correlation for all lag hours.

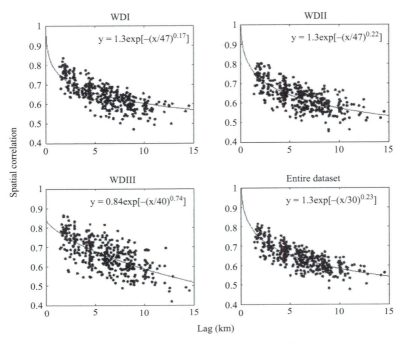

Figure 3 The relationship between spatial correlation coefficients of the radar rainfall and lag distances for three wind ranges and the entire dataset (Dai et al. 2015).

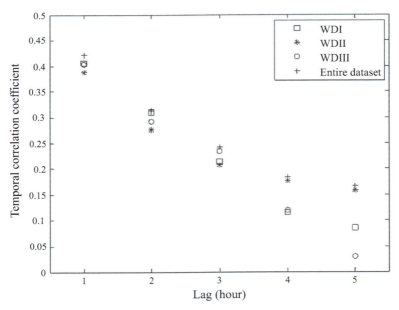

Figure 4 The relationship between temporal correlation coefficients of the radar rainfall and lag time for three wind ranges and the entire dataset (Dai et al. 2015).

Radar Rainfall Uncertainty Bands

The simulated uncertainty bands of WDI 1-4, WDII 1-4 and WDIII 1-4 are shown in Fig. 5 for radar Grid 1. Radar grid 1 locates in the northwest part of the catchment (see Fig. 1). In the figures, 500 ensemble members in each time step and the values of 5th to 95th percentile values are used to produce the uncertainty bands. The spread of uncertainty bands of WDIII events seems to be greater than ones generated under WDI and WDII conditions, indicating the wind field may affect the behaviors of the uncertainty bands.

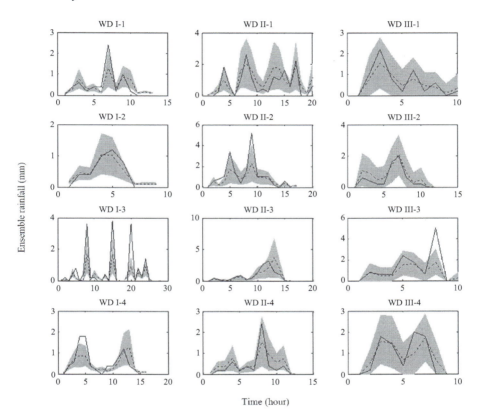

Figure 5 Uncertainty bands generated by 500 ensemble realizations using the t-copula for 12 events for the radar pixel 1. Each wind domain corresponds to four events. The solid line represents the reference gauge rainfall and the shaded area is the simulated ensemble members with the dashed line as the deterministic fields (Dai et al. 2015).

Impact of Synoptic Regimes on Radar Rainfall Uncertainty

Uncertainty bands in Fig. 5 reveal the different behaviors of radar rainfall uncertainty in different wind conditions. To better investigate how the uncertainty bands vary with wind field, the dispersion and bias of generated ensemble members for 30 events (10 events for each wind domain) are calculated using Equations (5) and (6), which are listed in Table 4. The dispersion statistics grow remarkably with the increase of wind speed. In other words, the radar rainfall uncertainty for strong wind condition is larger than ones during weak wind condition.

TABLE 4 Statistics of dispersion and ensemble bias of simulated uncertainty bands. To better show the results, the event IDs only list the index.

Event ID	Dispersion			Ensemble bias		
	WD I	**WD II**	**WD III**	**WD I**	**WD II**	**WD III**
1	0.69	1.64	1.85	0.33	0.68	0.75
2	0.91	1.93	2.28	0.21	0.73	0.86
3	0.73	2.10	2.46	0.44	0.66	0.56
4	0.85	1.21	2.45	0.33	0.40	0.84
5	0.86	1.29	1.52	0.40	0.49	0.69
6	1.18	1.48	2.04	0.55	0.59	0.87
7	1.17	1.27	2.03	0.57	0.41	0.60
8	1.11	0.96	1.77	0.44	0.33	0.64
9	1.34	2.12	1.82	0.84	0.77	0.83
10	0.66	1.87	1.99	0.41	0.86	0.72

Potential Applications of Ensemble Generator

The outcome of the ensemble generator can be used as a direct input into the hydrologic forecasting system, which provides the possibilities to examine the propagation of rainfall uncertainty in these systems. Cunha et al. (2012) diagnosed the uncertainty propagation in which the uncertainties of radar-rainfall affected peak flow simulation uncertainties through inputting the generated rainfalls into a well-established hydrological model. Once we have obtained a clear understanding of input uncertainty, we can concentrate on the parameters and structural uncertainties of hydrologic models. The behavior of the outcomes of these systems can inversely verify the performance of the proposed scheme. If we want to do this, it is obviously a premise to relax the influence of the uncertainty from those systems themselves. As the ensemble generator can generate a time series of ensemble rainfall, it is capable of being incorporated in the implementation of a real-time hydrologic system.

Furthermore, it is an elegant solution to integrate the ensemble generator with the data assimilation system for rainfall forecast. The multivariate distribution of random error derived by the proposed scheme is of great importance in the assimilation processing. For instance, the optimal performance can be achieved for rainfall forecasts by blending the radar rainfall with a NWP downscaling scheme if the uncertainties of RR and NWP can be modeled. It is worth remarking that the observed error is assumed to be Gaussian in most data assimilation schemes, such as 3D/4D Var and the ensemble Kalman filter.

Based on the proposed scheme, it is feasible to design a radar-based probabilistic forecasting chain by combining the ensemble generator with the radar rainfall forecasting model or numerical weather prediction model. Then we can describe the probable uncertainties of the predicted radar-rainfall. A simple method is to use the generated ensemble members of the estimated rainfall as the input to the rainfall forecasting model. Each generated member will produce a time series of forecast results. For instance, Liechti et al. (2013) implemented a probabilistic forecasting system, where the COSMO-2 is initialized with 25 different initial rainfalls. However, this scheme only accounts for the uncertainties of the radar estimated rainfall, without considering the errors of the forecast models. Instead, we can build the error model for each forecast lead time using a similar scheme proposed herein. For example, we can establish an empirical uncertainty model using all the predicted results of the one hour lead time based on the long-term historic data.

○ CONCLUSIONS

The goal of this study is to calculate the probability distribution of 'true' rainfall once given a radar estimate, which reveals the degree of confidence in the radar products. A clear knowledge of input radar rainfall uncertainty will support the analysis of the hydrological model uncertainty. A formulated radar rainfall uncertainty model is first presented. Then a scheme named Multivariate Distributed Ensemble Generator (MDEG) is proposed for generating ensemble rainfall fields. The wind field is introduced into the MDEG and the radar measured uncertainty is quantified under different wind conditions. Other factors such as seasons and storm type are also considered in this study.

In summary, there are five key findings of this chapter: 1) the uncertainty model of radar rainfall based on the conditional GR|RR distribution can perform well in representing the marginal distribution and spatial and temporal dependence of radar rainfall uncertainty; 2) The random component of radar rainfall error is correlated in space and time and the relationship between the correlation and lag distance can be parameterized as an exponential function; 3) The fair agreement of the spatial and temporal rank correlation coefficients between the original observed data and the theoretically derived ones by the copula suggests that the copula is able to accurately simulate the spatial and temporal dependence for the random error of RR; 4) The ensemble generator based on the copula and autoregressive filter can represent the radar rainfall uncertainty quite well. This generator can fully model the marginal distribution, spatial dependence and temporal dependence for radar rainfall; 5) the wind speed strongly affects the behavior of the radar quality. The radar rainfall uncertainty model established under different wind conditions should be more realistic in representing the radar rainfall uncertainty.

○ REFERENCES

Ciach, G.J., Krajewski, W.F. and Villarini, G. 2007. Product-error-driven uncertainty model for probabilistic quantitative precipitation estimation with NEXRAD data. Journal of Hydrometeorology 8: 1325-1347.

Collier, C. 1986. Accuracy of rainfall estimates by radar, Part I: Calibration by telemetering raingauges. Journal of Hydrology 83: 207-223.

Cunha, L.K., Mandapaka, P.V., Krajewski, W.F., Mantilla, R. and Bradley, A.A. 2012. Impact of radar – rainfall error structure on estimated flood magnitude across scales: an investigation based on a parsimonious distributed hydrological model. Water Resources Research 48(10): W10515.

Dai, Q., Han, D., Rico-Ramirez, M. and Srivastava, P.K. 2014a. Multivariate distributed ensemble generator: A new scheme for ensemble radar precipitation estimation over temperate maritime climate. Journal of Hydrology 511: 17-27.

Dai, Q., Han, D., Rico-Ramirez, M.A. and Islam, T. 2014b. Modelling radar-rainfall estimation uncertainties using elliptical and Archimedean copulas with different marginal distributions. Hydrological Sciences Journal 59: 1992-2008.

Dai, Q., Han, D., Rico-Ramirez, M.A., Zhuo, L., Nanding, N. and Islam, T. 2015. Radar rainfall uncertainty modelling influenced by wind. Hydrological Processes 29: 1704-1716.

Germann, U., Berenguer, M., Sempere-Torres, D. and Zappa, M. 2009. REAL—Ensemble radar precipitation estimation for hydrology in a mountainous region. Quarterly Journal of the Royal Meteorological Society 135: 445-456.

Godambe, V. 1960. An optimum property of regular maximum likelihood estimation. The Annals of Mathematical Statistics 31: 1208-1211.

Godambe, V. 1976. Conditional likelihood and unconditional optimum estimating equations. Biometrika 63: 277-284.

Harrold, T., English, E. and Nicholass, C. 1974. The accuracy of radar – derived rainfall measurements in hilly terrain. Quarterly Journal of the Royal Meteorological Society 100: 331-350.

He, X., Refsgaard, J.C., Sonnenborg, T.O., Vejen, F. and Jensen, K.H. 2011. Statistical analysis of the impact of radar rainfall uncertainties on water resources modeling. Water Resources Research 47(9): 1-17.

Joe, H. 1997. Multivariate Models and Dependence Concepts. Chapman Hall, London.

Kitchen, M., Brown, R. and Davies, A. 1994. Real – time correction of weather radar data for the effects of bright band, range and orographic growth in widespread precipitation. Quarterly Journal of the Royal Meteorological Society 120: 1231-1254.

Liechti, K., Panziera, L., Germann, U. and Zappa, M. Flash-flood early warning using weather radar data: from nowcasting to forecasting. EGU General Assembly Conference Abstracts, 2013. 8490.

Malevergne, Y. and Sornette, D. 2003. Testing the Gaussian copula hypothesis for financial assets dependences. Quantitative Finance 3: 231-250.

Moore, R., Jones, D., Cox, D. and Isham, V. 2000. Design of the HYREX raingauge network. Hydrology and Earth System Sciences Discussions 4: 521-530.

Priestley, M. 1981. Spectral Analysis and Time Series. Academic Press, London.

Roberts, A., Cluckie, I., Gray, L., Griffith, R., Lane, A., Moore, R. and Pedder, M. 2000. Appendix: Data management and data archive for the HYREX Programme. Hydrology and Earth System Sciences Discussions 4: 669-679.

Seed, A. 2003. A dynamic and spatial scaling approach to advection forecasting. Journal of Applied Meteorology 42: 381-388.

Villarini, G. and Krajewski, W. F. 2010. Review of the different sources of uncertainty in single polarization radar-based estimates of rainfall. Surveys in Geophysics 31: 107-129.

Wheater, H., Isham, V., Cox, D., Chandler, R., Kakou, A., Northrop, P., Oh, L., Onof, C. and Rodriguez-Iturbe, I. 1999. Spatial-temporal rainfall fields: modelling and statistical aspects. Hydrology and Earth System Sciences 4: 581-601.

Wood, S., Jones, D. and Moore, R. 2000. Accuracy of rainfall measurement for scales of hydrological interest. Hydrology and Earth System Sciences Discussions 4: 531-543.

Xu, J.J. 1996. Statistical modelling and inference for multivariate and longitudinal discrete response data. Ph.D. thesis, Department of Statistics, University of British Columbia.

17
CHAPTER

◇◇

Soil Moisture Retrieval from Bistatic Scatterometer Measurements using Fuzzy Logic System

Dileep Kumar Gupta,[1] Rajendra Prasad,[1,] Prashant K. Srivastava,[2,3,4] Tanvir Islam[5,6] and Manika Gupta[2]*

ABSTRACT

Many hydrological processes are strongly dependent on the soil moisture content on top of earth surfaces. In the present study, the fuzzy logic algorithms are used for the retrieval of soil moisture using bistatic scatterometer data. For this purpose, the bistatic scatterometer measurements are performed for the rough soil surface at the incidence angles 20° to 70° steps of 5° for HH- and VV-polarization at different soil moisture contents. The linear regression analysis is performed between observed soil moisture contents and bistatic scattering coefficients for selecting the suitable incidence angle for soil moisture retrieval. The bistatic scattering coefficients at 25° incidence angle are selected as the input data sets for the calibration and validation of fuzzy logic models. The performances of fuzzy models are assessed in terms of statistical performance indices %Bias, root mean squared error (RMSE) and Nash-Sutcliffe Efficiency (NSE). The analysis of the above mentioned indices indicates that the fuzzy model at VV-polarization is found better than HH-polarization.

KEYWORDS: Soil moisture, Scatterometer, Fuzzy logic, Co-polarization.

[1] Department of Physics, Indian Institute of Technology (B.H.U.), Varanasi, India.
[2] Hydrological Sciences, NASA Goddard Space Flight Center, Greenbelt, Maryland, USA.
[3] Earth System Science Interdisciplinary Center, University of Maryland, Maryland, USA.
[4] Institute of Environment and Sustainable Development, Banaras Hindu University, Varanasi, India.
[5] NASA Jet Propulsion Laboratory, Pasadena, CA, USA.
[6] California Institute of Technology, Pasadena, CA, USA.
* Corresponding author: rprasad1@rediffmail.com

○ INTRODUCTION

Soil moisture is a key variable on the interface between atmosphere and land surfaces, plays an important role in climate changes like exchange of water and heat energy fluxes and influences the hydrological cycle (Srivastava 2013; Srivastava et al. 2013a). The soil moisture at the top of land surfaces plays an important role in the development of weather patterns and the production of precipitation (Srivastava et al. 2013b). The soil moisture directly evaporates from the land surfaces due to the sun radiation or the vegetation that absorbs soil moisture through the roots and the radiation energy that induces transpiration. It illustrates that the knowledge about the dynamics of soil moisture depends on our understanding of the global energy and water cycle (Srivastava et al. 2013). Therefore a systematic technique to measure the spatial and temporal distribution of soil moisture with reasonable accuracy at local, regional and global levels is strongly required (Arii 2009).

The microwave remote sensing technique is more popular for the retrieval and monitoring of land surface parameters due to its weather independent imaging capability (Jackson and Schmugge 1989; Kawanishi et al. 2003). The retrieval of soil moisture through microwave remote sensing depends on the dielectric and geometrical properties of the land surfaces and it is also depends upon the system parameters like frequency, polarization and incidence angle (Kerr et al. 2012; Srivastava, et al. 2014). The microwave is more sensitive towards large dissimilarity between the dielectric constant of liquid water (~80) and dry soil (~4) (Dobson et al. 1985; Hallikainen et al. 1985; Srivastava et al. 2014; Wang and Schmugge 1980).

In many cases, the monostatic system has been used for retrieving the soil moisture (Baghdadi et al. 2002; Ceraldi et al. 2005; Chauhan 1997; Du et al. 2000; Dubois et al. 1995; Engman and Chauhan 1995; Njoku and Li 1999; Oh et al. 1992; Paloscia et al. 2008; Schmugge et al. 1986; Singh 2005; Ulaby et al. 1982; Wang et al. 1983) and the lesser studies have been established for the bistatic system (Brogioni et al. 2010; Khadhra et al. 2012; Pierdicca et al. 2008; Singh et al. 1996). At present, more study on bistatic scatterometer is needed to enhance the knowledge about the bistatic system for the retrieving of soil surface parameters. The German Aerospace Centre (DLR) in association with Astrium GmbH launched the TanDem-X satellite on 21 June 2010, which is the second satellite in the series of TerraSAR-X and it provides the bistatic data in X-band. The understanding of scattering mechanism in relation to soil moisture from natural soil surfaces at various microwave frequencies is very difficult to understand. During the last decade, many empirical, semi-empirical and physical models have been developed for retrieving the soil surface parameters (Fung et al. 1992; Oh et al. 1992; Saleh et al. 2006; Singh 2005; Singh et al.1996; Song et al. 2013; Srivastava et al. 2014). It is a worthy pursuit to develop the simplest models for retrieving the soil moisture using radar data. Nowadays, several researchers have studied an artificial neural network (ANN) as the estimator to estimate the soil moisture and other soil surface parameters using radar data (Chai et al. 2009; Del Frate et al. 2003; Dharanibai and Alex 2009; Jiang and Cotton 2004). Few researchers have studied fuzzy logic for the estimation of soil moisture using radar data (Lakhankar et al. 2006). Hence, the enhancement of knowledge about the fuzzy logic for the estimation of soil moisture using radar data is greatly required.

The objectives of the present investigation are to study the bistatic scattering coefficients for the estimation of soil moisture of soil rough surfaces using fuzzy logic and bistatic scatterometer data at X-band. Linear regression analysis is carried out between scattering coefficients and soil moisture to find the suitable incidence angle for the estimation of soil moisture at HH- and VV-polarization. A fuzzy inference system is calibrated using input data sets (bistatic scattering coefficients) and output data sets (soil moisture content in rough soil surface). The calibrated fuzzy inference system is used for the estimation of soil moisture of rough soil surfaces.

○ METHODS AND OBSERVATIONS

Soil Moisture Measurement

An outdoor soil test bed of bare soil surface (4 m × 4 m) is specially prepared and flooded with water for 20 to 24 hours to have a large range of soil moisture contents before making observations. The soil moisture content is usually measured in two terms namely gravimetric soil moisture content and volumetric soil moisture content. In this study, the gravimetric moisture content of the soil surface is taken. The gravimetric soil moisture content is defined as the ratio of the weight of water present in the soil to the weight of dry soil. It is expressed as a percentage of soil moisture content. Five randomly soil samples are collected in an aluminum soil container up to the depth of 5 cm from the soil surface. These soil samples were dried in an oven at 110°C for 20 hours. The samples were weighted before and after drying to compute the gravimetric moisture content. The average of the gravimetric moisture content of all the five soil samples were taken to calculate the percentage of soil moisture content of the soil surface. Fig. 1 shows the instruments required for the measurement of gravimetric soil moisture. The percentage of gravimetric soil moisture content is given as:

$$Mg(\%) = \frac{\text{Weight of wet soil sample} - \text{weight of dry soil sample}}{\text{weight of dry soil sample}} \times 100 \qquad (1)$$

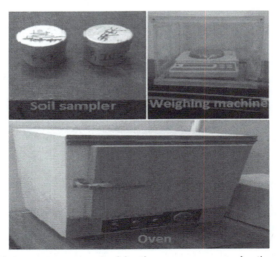

Figure 1 Instruments required for the measurement of soil moisture.

Measurement of Surface Roughness

The pin profiler is used to measure the soil surface roughness. The length of the profiler is 1 meter. The pin profiler contains 100 straight spokes of metal. The distance between each spoke is 1 cm such that the pin profiler has a total of 100 spokes. The pin profiler is inserted in the rough soil surface at the level of the maximum depth of the soil surface. The variation according to the tip of all spokes of the pin profiler creates a spectrum according to the roughness of the soil surface. This spectrum draws on the white paper and takes the photograph of this spectrum. This spectrum digitizes it into 1000 data points using a computer program. The vertical height from the mean of roughness spectrum is generated for all the 1000 data sets. The generated data sets are simulated in the computer to find the RMS height, correlation length and autocorrelation function. The RMS height and auto correlation length of surface roughness is found 1.61 cm and 11.69 cm respectively. Fig. 2 shows the graphical representation of rough surface spectrum, auto correlation function and the comparison of the auto correlation function with the Gaussian auto correlation function and exponential auto correlation function. The auto correlation function is at par with the Gaussian auto correlation function. It means, the surface can be considered as the Gaussian surface.

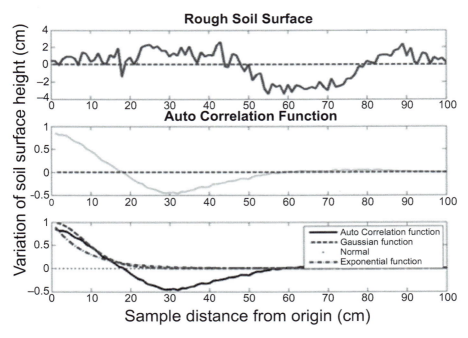

Figure 2 The plot for spectrum of surface roughness and auto correlation function (RMS height (σ) = 1.61 cm and correlation length (l) = 11.69 cm).

Bistatic Scatterometer Measurments and Computation of Scattering Coefficients

Fig. 3 and Table 1 shows the photograph and specifications of the experimental setup of the bistatic scatterometer system respectively (Ulaby et al. 1982). The bistatics catterometer system has a separate transmitter and receiver. The transmitter consists of a pyramidal dual polarized X-band horn antenna, a waveguide to N-female coaxial adaptor and PSG high power signal generator (E8257D, 10 MHz to 20 GHz). The receiver of bistatic scatterometer consists of apyramidal dual polarized X-band horn antenna, a waveguide to N-female coaxial adaptor, EPM- P series power meter (E4416A) and peak and average power sensor (E9327A, 50 MHz – 18 GHz).

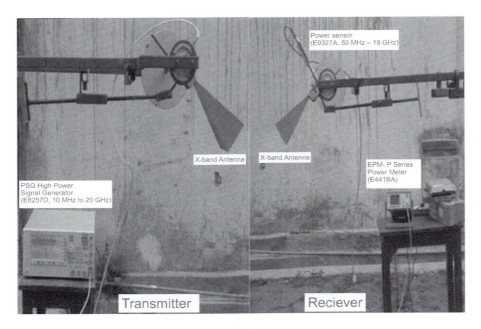

Figure 3 Photograph of bistatic scatterometer.

The bistatic scatterometer is designed to take the measurements of scattered power from the rough soil surface at different soil moisture contents in the angular range of incidence angle 20° to 70° steps of 5° for HH- and VV-polarizations at X-band. The bistatic scatterometer has the facility to change the height and incidence angle of the antennas. A laser pointer is used to maintain the focus of the antenna beam at the centre of test soil surface by adjusting the height of the antenna at different incidence angles. The incidence angle and height of the antenna are measured by the pointer provided on the circular and linear scale respectively.

TABLE 1 Specifications of bistatic scatterometer system.

RF generator		E8257D, PSG high power signal generator, 10 MHz to 20 GHz (agilent technologies)
Power meter		E4416A, EPM-P series power meter, 10 MHz to 20 GHz (agilent technologies)
Power sensor		Peak and average power sensor (E9327A, 50 MHz – 18 GHz)
Frequency (GHz)		10 ± 0.05 (X – Band)
Beam	E plane (°)	17.3118
Width	H plane (°)	19.5982
Antenna gain (dB)		20
Cross-polarization isolation (dB)		40
Polarization modes		HH– and VV–
Antenna type		Dual-polarized pyramidal horn
Calibration accuracy		1 dB
Incidence angle (°)		20° (nadir) – 70°

Fig. 4 shows the aluminium sheet used for the calibration of the bistatic scatterometer system (Curie 1989). This system has the capability to measure the reflected and transmitted power for the angular range 20° to 70° by steps 5° at HH- and VV-polarizations for X-band (10 GHz). If λ is the wavelength of the transmitted wave, P_t is the transmitted power G_r and G_t are the gain of receiving and the transmitting antennas, R_1 and R_2 are the distance of the transmitting and receiving antenna from the centre of illuminated area; then the power received at the receiver due to the perfectly conducting flat aluminum sheet as the reflecting surface is given by

$$P_r^{Al} = \frac{P_t G_r G_t \lambda^2}{(4\pi)^2} (R_1 + R_2)^2 \tag{2}$$

If the reflectivity of a reflecting target is R_0 then the received power can be expressed as

$$P_r = \frac{P_t G_r G_t \lambda^2 |R_0|^2}{(4\pi)^2} (R_1 + R_2)^2 \tag{3}$$

The reflectivity (r) of the target may be obtained by the equation

$$r = |R_0|^2$$
$$= \frac{P_r}{P_r^{Al}} \tag{4}$$

For the Fraunhofer zone observation, the range ($R = R_1 = R_2$) can be taken as large enough so that the R could be considered constant over A_0 (the surface); the radar equation (Ulaby et al. 1982) reduces

$$P_r = P_t \lambda^2 \oint_{A_0} \frac{G_r G_t \sigma^0 ds}{(4\pi)^2 R^4} \tag{5}$$

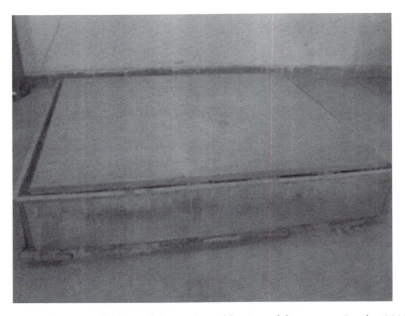

Figure 4 Aluminum sheet used during the calibration of the system (Pandey 2011).

For the average measurement of scattering coefficient of the target such as crop and soil in our case, the scatterometer system was calibrated for the target of the known radar cross section.

The power received from a standard target i.e. perfectly flat and smooth aluminum plate is written as

$$P_r^{std} = \frac{P_t G_{tm} G_{rm} \lambda^2}{(4\pi)^2}(2R)^2 \tag{6}$$

where G_{tm} and G_{rm} represent the maximum gain of the transmitting and receiving antennas, respectively.

From equations (6) and (7), we get

$$\frac{P_r}{P_r^{std}} = \frac{I}{\pi R^2} \tag{7}$$

where

$$I = \oint_{A_0} \sigma^0 G_{tn} G_{rn}\, ds \tag{8}$$

and
$$G_{tn} = \frac{G_t}{G_{tm}}, \; G_{tn} = \frac{G_t}{G_{tm}}$$

Assuming scattering coefficient constant over a 3 dB band width of the antenna beam, we have

$$\sigma^0 = \pi R^2 \frac{P_r}{P_r^{Std}} \oint_{A_0} G_{tn} G_{rn} \, ds \tag{9}$$

From equation (5) and (9), we have

$$\sigma^0 = \pi R^2 \frac{|R_0|^2}{I_0} \tag{10}$$

where I_0 is the illuminated area of the target. An antenna beam falls on the target surface in the form of an ellipse. Its minor axis and major axis are given by

$$\text{Minoraxis} = 2R^2 \tan\left(\frac{\phi_{az}}{2}\right) \tag{11}$$

$$\text{Majoraxis} = R^2 \sin\left(\frac{\phi_{el}}{2}\right)\left[\sec\left(\theta - \frac{\phi_{el}}{2}\right) + \sec\left(\theta - \frac{\phi_{el}}{2}\right)\right] \tag{12}$$

where ϕ_{el}, ϕ_{az} and θ are elevation, azimuth and look angle of antennas respectively. Therefore, the illuminated area is given by

$$I_0 = \frac{\pi}{2} R^2 \tan\left(\frac{\phi_{az}}{2}\right) \sin\left(\frac{\phi_{el}}{2}\right)\left[\sec\left(\theta - \frac{\phi_{el}}{2}\right) + \sec\left(\theta + \frac{\phi_{el}}{2}\right)\right] \tag{13}$$

Now, substituting the value of I_0 in equation (10), the value of scattering coefficient is obtained as

$$\sigma^0 = \frac{2 |R_0|^2 \cot\left(\frac{\phi_{az}}{2}\right)\text{cosec}\left(\frac{\phi_{el}}{2}\right)}{\sec\left(\theta - \frac{\phi_{el}}{2}\right) + \sec\left(\theta + \frac{\phi_{el}}{2}\right)} \tag{14}$$

Therefore, the scattering coefficient in unit dB can be written as

$$\sigma^0(\text{dB}) = 10 \log_{10}\left[\frac{2 |R_0|^2 \cot\left(\frac{\phi_{az}}{2}\right)\text{cosec}\left(\frac{\phi_{el}}{2}\right)}{\sec\left(\theta - \frac{\phi_{el}}{2}\right) + \sec\left(\theta + \frac{\phi_{el}}{2}\right)}\right] \tag{15}$$

Therefore, knowing the value of the elevation, azimuth, look angle of an antenna and the reflectivity of the target, we can compute the scattering coefficient of the target by equation 15.

○ FUZZY LOGIC FOR ESTIMATION OF SOIL MOISTURE ═══════

Calibration and Validation Data Sets of Fuzzy Logic

The linear regression analysis is done for selecting the suitable incidence angle at HH- and VV-polarization. The higher value of coefficients of determination (R^2) is found at 25° incidence angle for both polarizations. In the present study, the bistatic scatterometer measurements are carried out for four soil moisture conditions in one test soil surface. The fuzzy logic modeling required more data sets for the calibration and validation of model. It is required to generate more data sets. For this purpose, the observed bistatic scattering coefficients and soil moisture are interpolated into 68 data sets at 25° incidence angle for HH- and VV-polarization. The Cubic Spline interpolation method is used to perform the interpolation of data sets. The interpolated data sets are used for the calibration and validation of the fuzzy logic model. 75% of the data sets are used for the calibration while the remaining 25% data sets are kept separately for the validation purposes of the fuzzy logic model.

The interpolated HH- or VV-polarized bistatic scattering coefficients are used as the input data sets while the interpolated soil moisture contents are used as the output data sets for the fuzzy logic model. Two fuzzy logic models are calibrated for the validation of soil moisture contents using HH- and VV-polarization.

Performance Indices

This study used some performance indices for the estimation of soil moisture content from bare and rough soil surfaces using bistatic scatterometer data in a specular direction. The performance indices namely, %Bias, root mean squared error (RMSE) and Nash-Sutcliffe Efficiency (NSE) are used.

The percentage bias (%Bias) measures the average tendency of the estimated values to be larger or smaller than their observed values. The optimum value of %Bias is 0.0 and the smaller values of %Bias indicate accurate model predictions.

$$\%\text{Bias} = 100 * \left[\frac{\Sigma(y_i - x_i)}{\Sigma x_i} \right] \tag{16}$$

Root-mean-square error (RMSE) is a frequently used to measure the differences between estimated values by a model or an estimator and the observed values.

$$\text{RMSE} = \sqrt{\frac{1}{n} \sum_{i=1}^{n} (y_i - x_i)^2} \tag{17}$$

where n is the number of observations.

The Nash-Sutcliffe Efficiency (NSE) is based on the sum of absolute squared differences between the estimated and observed values normalized by the variance of the observed values during the study. The NSE was calculated using the given formula, the observed values during the study. The NSE was calculated using the given formula,

$$\text{NSE} = 1 - \frac{\sum_{i=1}^{n}(y_i - x_i)^2}{\sum_{i=1}^{n}(x_i - \bar{x}_i)^2} \tag{18}$$

where n is the number of observations; x_i is the observed and y_i is the simulated variables.

Fuzzy Logic

Fuzzy logic is a systematic mathematical approach to deal with membership functions, fuzzy sets and fuzzy operators to form the many set of rules to establish the suitable relationship between given input and output. The fuzzy rules are deals with approximate, rather than fixed and exact reasoning. Fuzzy logic provides a better understanding of the things present in the world that surrounds us that is defined by a non-distinct boundary based upon vague, ambiguous, imprecise, noisy and missing input information (Priyono et al. 2012).

Matlab software has a fuzzy logic toolbox to generate the fuzzy inference system (FIS). The fuzzy inference system uses fuzzy set theory for the mapping of inputs to outputs. In the present study, the MATLAB® software based genfis2 algorithm is used for the subtractive clustering method to generate the Fuzzy Inference System (FIS). This Fuzzy inference system is a Sugeno type fuzzy model with a systematic approach for generating fuzzy rules from a given input-output data set (Sugeno and Kang 1988; Takagi and Sugeno 1985).

The subtractive clustering method (Chiu 1994) is used for grouping of the same categories of data sets and to identify the same categories of data sets from HH- and VV-polarized bistatic scattering coefficients along with soil moisture data sets. This is the unsupervised data clustering method. So the information about the number of clusters in the data sets don't need required priory. This unsupervised data clustering method is used as aone-pass algorithm for estimating the number of clusters and the cluster centers in a set of data. The subtractive clustering method assumes each data point is a potential cluster center and calculates the likelihood that each data point would define the cluster center, based on the density of surrounding data points. First consider all the data points with highest potential to be the first cluster and remove all the data points. The surrounding area of the cluster center is determined by the *radii* of the clusters. The next data cluster and its center location can be determined by the iterative process of all data within *radii* of the cluster center.

The fuzzy inference system is realized through data clustering for the retrieval of soil moisture of rough soil surfaces using bistatic scatterometer data. The retrieval of soil moisture is performed by the inputs (bistatic scattering coefficients at HH- and VV-polarization). For the fuzzy inference system, the bistatic scattering coefficients (HH- or VV-polarized) are considered as the input data while the soil moisture content is considered as the output data. Thus, two data sets are prepared to generate the fuzzy inference system at both polarizations (HH- and VV-).

The *subclust* function is inbuilt in *genfis2* function in the MATLAB for the subtractive clustering algorithm. The clustering of data sets helps to set the number of fuzzy rules. The number of clusters gives the same number of fuzzy rules. More accurately, it is categorizing the entire data sets into different clusters depending upon the variability of scattering coefficients at HH- and VV-polarization with the variation of the soil moisture in the rough soil surfaces. The small value of *radii* accounts for a greater number of clusters while the higher value of *radii* accounts for a lesser number of clusters.

The selection of optimum value of *radii* is an important parameter for getting the optimum retrieval accuracy of soil moisture using bistatic scatterometer data sets. The value of optimum radii is found by simulating the fuzzy model with different values of radii having range 0.1 to 1 steps of 0.05 and monitoring the error between observed and estimated values of soil moisture continuously. In this study, the value of *radii* is found 0.5 for the retrieval of soil moisture using both inputs data sets (scattering coefficients at HH- or VV-polarization). The values of number of clusters are found as 4 and 3 for HH- and VV-polarizations respectively. The values of each cluster center for the combination of input (bistatic scattering coefficients) and output (soil moisture) are found as 6.16, 8.23, 9.37, 7.11 and 14.16, 20.36, 22.9600, 17.76 at HH-polarization. For the VV-polarization, the values of each cluster center for the combination of input (bistatic scattering coefficients) and output (soil moisture) are found as 0.4990, 5.24, 8.81 and 14.16, 19.76, 22.76 respectively.

Figs. 5a-b show the plot between the observed and estimated soil moisture with 1:1 equi. line from rough soil surfaces using bistatic scatterometer data during calibration of fuzzy models at HH- and VV-polarization respectively. The 1 : 1 equi. line provides clear visibility for users to obtain the variation in estimated data sets from the observed data sets. The estimated soil moisture at HH- polarization has more variation than estimated soil moisture at VV- polarization from the observed soil moisture. The performance during the calibration and validation of fuzzy models for the estimation of soil moisture are evaluated in terms of %Bias, root mean square error (RMSE) and Nash-Sutcliffe Efficiency (NSE). Table 1 depicts the performance indices during calibration and validation of fuzzy models for the estimated soil moisture. The %Bias reveals the percentage of over/under estimated values by the models. The RMSE value measures the closeness of the estimated values to the observed values. The value of RMSE closer to zero indicates high reliability of estimated and observed values. The fuzzy models are underperformed during calibration at HH- and VV-polarization. The value of RMSE is found as 1.2861 and 1.1823 between observed and simulated soil moisture during calibration of fuzzy logic at HH- and VV-polarization. The calibration of fuzzy model is found better at VV-polarization than HH-polarization.

Figs. 6a-b show the plot between the observed and estimated soil moisture with 1:1 equi line from rough soil surfaces using bistatic scatterometer data during validation of fuzzy models at HH- and VV-polarization respectively. The fuzzy model shows the under-performance for validation at HH- and VV-polarization. The values of RMSE are found as 1.0581 and 0.9553 at HH- and VV-polarization respectively during validation of the fuzzy model. The values of NSE are found as 0.9271 and 0.9406 at HH- and VV-polarization respectively. According to the performances indices, the estimation of soil moisture from bare and rough soil surfaces using bistatic scatterometer data by fuzzy logic is found better for VV-polarization than HH-polarization.

TABLE 2 Statistical performance indicies between observed and estimated soil moisture during calibration and validation of fuzzy model.

	Calibration_HH	Validation_HH	Calibration_VV	Validation_VV
RMSE	1.2861	1.0581	1.1823	0.9553
Pbias	−3.42e-006	−0.7329	−1.62e-015	−0.7318
NSE	0.8925	0.9271	0.9091	0.9406

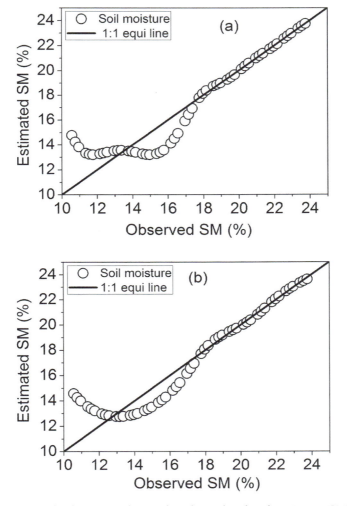

Figure 5 Scatter plot between observed and simulated soil moisture (SM) with 1:1 equi. line during calibration of fuzzy inference system (a) at HH- polarization (b) at VV-polarization.

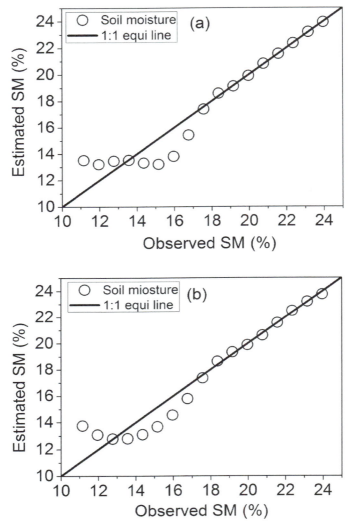

Figure 6 Scatter plot between observed and simulated soil moisture (SM) with 1:1 equi. line during validation of fuzzy inference system (a) at HH- polarization (b) at VV-polarization.

Figs. 7 and 8 show the input values (bistatic scattering coefficients) are coded in the form of fuzzy membership functions that lie between 0 and 1 at HH- and VV-polarization respectively for each cluster. The Gaussian membership function is used in this algorithm to represent the inputs. The fuzzy rules are generating using 51 calibration data sets and 17 data sets are used for the validation of fuzzy rules.

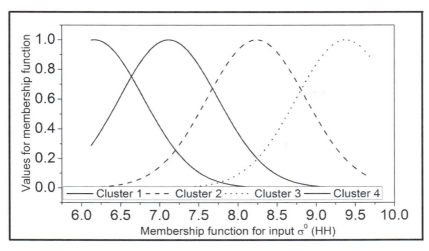

Figure 7   Relationship of bistatic scattering coefficients with membership values at HH-polarization for each cluster.

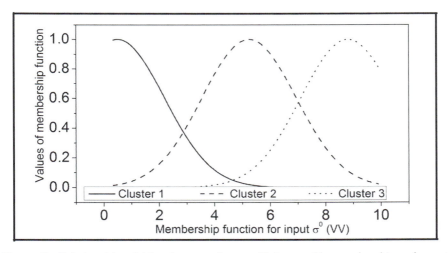

Figure 8   Relationship of bistatic scattering coefficients with membership values at VV-polarization for each cluster.

○ CONCLUSION

The angular variation of the scattering coefficient is found to increase with the soil moisture content. The suitable incidence angle for the estimation of soil moisture of the bare and rough soil surface was found as 25° for both HH- and VV-polarization. The performance of fuzzy model was found marginally better at VV-polarization than HH-polarization during the calibration of model. Similar findings are also found during the validation with higher performance of VV-polarization than HH-polarization. The values of NSE are found higher during calibration and validation of the fuzzy inference system for the retrieval of soil moisture using bistatic scattering coefficients at HH- and VV-polarizations which indicate that fuzzy logic can be used for the estimation of soil moisture using bistatic radar scatterometer measurements.

○ REFERENCES

Arii, M. 2009. Retrieval of soil moisture under vegetation using polarimetric Radar California: California Institute of Technology, Pasadena.

Baghdadi, N., King, C., Bourguignon, A. and Remond, A. 2002. Potential of ERS and Radarsat data for surface roughness monitoring over bare agricultural fields: Application to catchments in Northern France: International Journal of Remote Sensing 23: 3427-3442.

Brogioni, M., Pettinato, S., Macelloni, G., Paloscia, S., Pampaloni, P., Pierdicca, N. and Ticconi, F. 2010. Sensitivity of bistatic scattering to soil moisture and surface roughness of bare soils. International Journal of Remote Sensing 31: 4227-4255.

Ceraldi, E., Franceschetti, G., Iodice, A. and Riccio, D. 2005. Estimating the soil dielectric constant via scattering measurements along the specular direction. IEEE Transactions on Geoscience and Remote Sensing 43: 295-305.

Chai, S-S., Walker, J.P., Makarynskyy, O., Kuhn, M., Veenendaal, B. and West, G. 2009. Use of soil moisture variability in artificial neural network retrieval of soil moisture. Remote Sensing 2: 166-190.

Chauhan, N.S. 1997. Soil moisture estimation under a vegetation cover: Combined active passive microwave remote sensing approach International Journal of Remote Sensing 18: 1079-1097.

Chiu, S.L. 1994. Fuzzy model identification based on cluster estimation. Journal of intelligent and Fuzzy Systems 2: 267-278.

Curie NCW, G.W. 1989. Radar Reflectivity Measurement: Technique and Applications. Artech House, Norwood, MA.

Del Frate, F., Ferrazzoli, P. and Schiavon, G. 2003. Retrieving soil moisture and agricultural variables by microwave radiometry using neural networks. Remote Sensing of Environment. 84: 174-183.

Dharanibai, G. and Alex, Z. 2009. ANN technique for the evaluation of soil moisture over bare and vegetated fields from microwave radiometer data. Indian Journal of Radio Space. 38: 283-288.

Dobson, M.C., Ulaby, F.T., Hallikainen, M.T. and El-Rayes, M.A. 1985. Microwave dielectric behavior of wet soil-Part II: dielectric mixing models. IEEE Transactions on Geoscience and Remote Sensing 23: 35-46.

Du, Y., Ulaby, F.T. and Dobson, M.C. 2000. Sensitivity to soil moisture by active and passive microwave sensors. IEEE Transactions on Geoscience and Remote Sensing 38: 105-114.

Dubois, P.C., Van Zyl, J. and Engman, T. 1995. Measuring soil moisture with imaging radars. IEEE Transactions on Geoscience and Remote Sensing 33: 915-926.

Engman, E.T. and Chauhan, N. 1995. Status of microwave soil moisture measurements with remote sensing. Remote Sensing of Environment 51: 189-198.

Fung, A.K., Li, Z. and Chen, K. 1992. Backscattering from a randomly rough dielectric surface. IEEE Transactions on Geoscience and Remote Sensing 30: 356-369.

Hallikainen, M.T., Ulaby, F.T., Dobson, M.C., El-Rayes, M.A. and Lil-Kun, W. 1985. Microwave dielectric behavior of wet soil-Part 1: empirical models and experimental observations. IEEE Transactions on Geoscience and Remote Sensing 23: 25-34.

Jackson, T.J. and Schmugge, T.J. 1989. Passive microwave remote sensing system for soil moisture: Some supporting research. IEEE Transactions on Geoscience and Remote Sensing 27: 225-235.

Jiang, H. and Cotton, W.R. 2004. Soil moisture estimation using an artificial neural network: a feasibility study. Canadian Journal of Remote Sensing 30: 827-839.

Kawanishi, T., Sezai, T., Ito, Y., Imaoka, K., Takeshima, T., Ishido, Y., Shibata, A., Miura, M., Inahata, H. and Spencer, R.W. 2003. The advanced microwave scanning radiometer for the earth observing system (AMSR-E), NASDA's contribution to the EOS for global energy and water cycle studies. IEEE Transactions on Geoscience and Remote Sensing 41: 184-194.

Kerr, Y.H., Waldteufel, P., Richaume, P., Wigneron, J.-P., Ferrazzoli, P., Mahmoodi, A., Al Bitar, A., Cabot, F., Gruhier, C. and Juglea, S.E. 2012. The SMOS soil moisture retrieval algorithm. IEEE Transactions on Geoscience and Remote Sensing 50: 1384-1403.

Khadhra, K.B., Boerner, T., Hounam, D. and Chandra, M. 2012. Surface parameter estimation using bistatic polarimetric X-band measurements. Progress in Electromagnetic Research 39: 197-223.

Lakhankar, T., Ghedira, H. and Khanbilvardi, R. 2006. Neural Network and Fuzzy Logic for an Improved Soil Moisture Estimation. ASPRS, Annual Conference, Reno, Nevada.

Njoku, E.G. and Li, L. 1999. Retrieval of land surface parameters using passive microwave measurements at 6-18 GHz. IEEE Transactions on Geoscience and Remote Sensing 37: 79-93.

Oh, Y., Sarabandi, K. and Ulaby, F.T. 1992. An empirical model and an inversion technique for radar scattering from bare soil surfaces. IEEE Transactions on Geoscience and Remote Sensing 30: 370-381.

Paloscia, S., Pampaloni, P., Pettinato, S. and Santi, E. 2008. A comparison of algorithms for retrieving soil moisture from ENVISAT/ASAR images. IEEE Transactions on Geoscience and Remote Sensing 46: 3274-3284.

Pandey, A. 2011. Crop signature studies by microwave remote sensing using soft computing techniques. Varanasi: Indian Institute of Technology (BHU).

Pierdicca, N., Pulvirenti, L., Ticconi, F. and Brogioni, M. 2008. Radar bistatic configurations for soil moisture retrieval: a simulation study. IEEE Transactions on Geoscience and Remote Sensing 46: 3252-3264.

Priyono, A., Ridwan, M., Alias, A.J., Rahmat, R.A.O., Hassan, A. and Ali, M.A.M. 2012. Generation of fuzzy rules with subtractive clustering. Jurnal Teknologi 43: 143-153.

Saleh, K., Wigneron, J-P., de Rosnay, P., Calvet, J-C. and Kerr, Y. 2006. Semi-empirical regressions at L-band applied to surface soil moisture retrievals over grass. Remote Sens Environ 101: 415-426.

Schmugge, T., O'Neill, P.E. and Wang, J.R. 1986. Passive microwave soil moisture research. IEEE Transactions on Geoscience and Remote Sensing 24: 12-22.

Singh, D., Mukherjee, P., Sharma, S. and Singh, K. 1996. Effect of soil moisture and crop cover in remote sensing. Adv Space Res 18: 63-66.

Singh, D. 2005. A simplistic incidence angle approach to retrieve the soil moisture and surface roughness at X-band. IEEE Transactions on Geoscience and Remote Sensing 43: 2606-2611.

Song, K., Zhou, X. and Fan, Y. 2013. Algorithm for the retrieval of soil moisture from the radar backscattering coefficient. HKIE Transactions 20: 124-132.

Srivastava, P., O'Neill, P., Cosh, M., Kurum, M., Lang, R. and Joseph, A. 2014. Evaluation of dielectric mixing models for passive microwave soil moisture retrieval using data from ComRAD Ground-Based SMAP simulator. IEEE Journal of Selected Topics in Applied Earth Observations and Remote Sensing PP: 1-10.

Srivastava, P.K. 2013. Soil Moisture Estimation from SMOS Satellite and Mesoscale Model for Hydrological Applications. PhD Thesis, University of Bristol, Bristol, United Kingdom.

Srivastava, P.K., Han, D., Ramirez, M.R. and Islam, T. 2013a. Appraisal of SMOS soil moisture at a catchment scale in a temperate maritime climate. Journal of Hydrology 498: 292-304.

Srivastava, P.K., Han, D., Ramirez, M.R. and Islam, T. 2013b. Machine learning techniques for downscaling SMOS satellite soil moisture using MODIS land surface temperature for hydrological application. Water Resources Management 27: 3127-3144.

Srivastava, P.K., Han, D., Rico-Ramirez, M.A., Al-Shrafany, D. and Islam, T. 2013c. Data fusion techniques for improving soil moisture deficit using SMOS satellite and WRF-NOAH land surface model. Water resources management 27: 5069-5087.

Srivastava, P.K., Han, D., Rico-Ramirez, M.A., O'Neill, P., Islam, T. and Gupta, M. 2014. Assessment of SMOS soil moisture retrieval parameters using tau–omega algorithms for soil moisture deficit estimation. Journal of Hydrology 519, Part A: 574-587.

Sugeno, M. and Kang, G.T. 1988. Structure identification of fuzzy model. Fuzzy Sets and Systems 28: 15-33.

Takagi, T. and Sugeno, M. 1985. Fuzzy identification of systems and its applications to modeling and control. IEEE Transactions on Systems, Man and Cybernetics, SMC-15: 116-132.

Ulaby, F.T., Moore, R.K. and Fung, A.K. 1982. Microwave Remote Sensing Active and Passive-Volume II: Radar Remote Sensing and Surface Scattering and Emission Theory: Reading, MA: Addison-Wesley.

Wang, J.R. and Schmugge, T.J. 1980. An Empirical Model for the Complex Dielectric Permittivity of Soils as a Function of Water Content. IEEE Transactions on Geoscience and Remote Sensing GE-18: 288-295.

Wang, J.R., O'Neill, P.E., Jackson, T.J. and Engman, E.T. 1983. Multifrequency measurements of the effects of soil moisture, soil texture and surface roughness. IEEE Transactions on Geoscience and Remote Sensing GE21(1): 44-51.

Section V

Challenges in Geospatial Technology for Water Resources Development

18

CHAPTER

◇◇◇

Challenges in Geospatial Technology for Water Resources Development

Prashant K. Srivastava,[1,2,3,*] and *Aradhana Yaduvanshi*[4]

ABSTRACT

Geospatial technology has revolutionized water resources development and research by expanding our knowledge of terrestrial environment system and processes. However, geospatial techniques cannot substitute ground-based methods towards providing high quality data at a point scale. The high superiority lies in mapping conditions at regional, continental and even global scales and on a recurring basis. This chapter discusses the importance and value of Remote Sensing and Geographical Information System (GIS) and points out the challenges in this field that should be taken care of by researchers, policy makers and practitioners in order to view the technology in a proper perspective.

KEYWORDS: Geospatial Technology, Water Resources, Challenges, Remote Sensing, Geographical Information System.

○ INTRODUCTION

Earth's physical environment and resources undergo certain, inevitable and virtually constant changes that often seriously affect or even threaten the well being of humans. A growing population with limited water resources poses a competing environment for different sectors such as industrial, agricultural and domestic needs for sustenance (Tilman et al. 2002). Water being a physiological necessity and its assessment, conservation and management has great concern for all those who

[1] Hydrological Sciences, NASA Goddard Space Flight Center, Greenbelt, Maryland, USA.
[2] Earth System Science Interdisciplinary Center, University of Maryland, Maryland, USA.
[3] Institute of Environment and Sustainable Development, Banaras Hindu University, Varanasi, India.
[4] Center of Excellence in Climatology, Birla Institute of Technology, Mesra, Ranchi, India.
* Corresponding author: prashant.just@gmail.com

manage, facilitate and utilize this important natural resource (Patel et al. 2012a; Patel et al. 2012b). Frequent occurrences of extreme events like floods and drought in the face of global climate change is the firm illustration of fundamental changes that altering water resources world wide, including India (Patel & Srivastava 2013, 2014). It calls for great challenges in the best use of available water resources through surface water capture and storage, inter basin transfer, ground water exploitation and watershed managment (Patel et al. 2012b; Yadav et al. 2014). To achieve maximum water use efficiency and to cope up with varying water resources availability conditions, real time information on various aspects, which control and influence the supply & utilization regimes are needed (Reynolds et al. 2000; Ebert et al. 2007).

The geospatial techniques are promising tools to aid water management decisions (Gupta & Srivastava 2010; Srivastava et al. 2011; Srivastava et al. 2012). Scientific efforts of water management are largely supported by geospatial techniques involving systematic approaches of real time information on different aspects of utilizations (Yaduvanshi et al. 2015). Scientists have developed and employed various methods for obtaining information or data about the changing environment, but nothing has been as revolutionary as geospatial technologies. The argument for Geospatial based techniques includes the fact that it gathers data in a cheaper, timely and more efficient way than conventional approaches (Okoye & Koeln 2003). The geospatial data sets which generally we gather are suitable for automated analysis in microcomputers.

Geospatial technologies enable scientist to obtain information about a material object, conditions of a geographic space, or an environmental phenomenon, through variety of approaches (Okoye & Koeln 2003). Thus, the geospatial technology are effective tools not only for collection, storage, management and retrieval of a multitude of spatial and non-spatial data, but also for spatial analysis and integration of these data to derive useful outputs and modeling (Gupta & Srivastava 2010; Islam et al. 2012; Srivastava et al. 2013b). Some of the variables used in developing geospatial technology can be retrieved from remote sensing images. However, space borne sensors estimate environmental variables quasiquantitatively and the process of converting and analysing the signals to the actual parameter are often fraught with challenges (Okoye & Koeln 2003). This chapter discusses the importance and value of geospatial technology and points out the weaknesses which are needed to be resolved in near future. It will also focus on some limitations and drawbacks of geospatial technology that researchers, policy makers and practitioners should be taken care with in order to view the technology in a proper perspective.

○ COMMON CHALLENGES IN REMOTE SENSING FOR WATER RESOURCES

Day-by-day the Earth remote sensing data retrieval, processing, distribution and application becomes more challenging and hence desires a more detailed scrutiny on a regular basis. Huge volumes of remote sensing data will be produced as the number of earth observing satellites are increasing day by day and there are many more coming in the near future (Srivastava et al. 2014d). Classical data processing and distribution methods generally suffer from poor performances and thus make them inadmissible for numerous end users and real time forecasting which require high performance spatial datasets (Dai et al. 2014). To assess local or global changes, both from natural and manmade influences for practical applications the increasing quality, quantity and duration of these observation are critically important. More detailed analysis and efficient datasets may help make the satellite data usage more reliable. One of the challenges is maintaining the high volumes of satellite datasets and transferring them into a form which is conveniently used by nearly all scientific communities again by utilising less space.

Custom-made satellite remote sensing data with an interactive graphical user interface on a GIS/Web-GIS compatible format is important for many users. End users need to obtain Earth observing remote sensing data in more useful forms. On the other hand, more widely spread Earth observing remote sensing data in different formats through diversified protocols will result in better management of future Earth observing satellite systems. Further, a proper analysis of remotely sensed data cannot be done without ground truthing. The two most important phases such as calibration (developing training areas) and validation (accuracy assessment) are needed to be done with reference dataset. Hence, although satellite systems have unprecedented and unparalleled capabilities to gather valuable information, the utility of many remotely sensed data and the success of most environmental investigations, may still rest significantly on the availability of quality data from non-remote sensing sources.

Being capable of providing unrestricted information with spatio-temporal revisits, remote sensing has a strong technological foundation, aiming at developing non-contact sensor and processing systems to gather reliable information about the Earth and other physical objects. Accurate, current and high resolution remotely sensed, geophysical information data makes the future of remote sensing accurate in many fields of need. Easy and quick import and analysis of data in non-specialized software packages is the dream of professionals in this field.

Challenges facing terrestrial and atmospheric remote sensing are unlimited as the different topography behaves differently with particular land surface dynamics and meteorological conditions across the globe at various spatial and temporal scales. Ultimate goals are to be able to make precise estimates of geophysical variables, with the intention of either advancing fundamental knowledge through development of empirical relationships and/or theoretical models or making predictions across time and spatial boundaries. From literature, several challenges are found in moving towards a better water resources development such as 1) Challenges in monitoring soil, snow and vegetation, 2) Challenges in soil moisture estimation, 3) Challenges in monitoring Evapotranspiration and energy fluxes, 4) Uncertainties in retrieval algorithms, 5) Bias associated with instruments and 6) Short spans of satellite data.

○ CHALLENGES IN MONITORING PRECIPITATION, SNOW AND VEGETATION

Space technology plays a significant role in supporting and evolving water management practices. Optimum planning of water resources projects can be achieved by combining conventional ground measurements along with remote sensing techniques. Methods have been used to quantify the components of the water and energy balance equations using remote sensing methods which includes the land surface temperature, near surface soil moisture, snow cover and vegetation.

Precipitation, particularly in the form of snow, strongly influences regional land-atmosphere interactions, ultimately impacting the global climate system. The information of the snow cover is crucial for hydrological and climate studies as well as required for radiative transfer models. For this reason, a number of satellite sensors are employed to provide snow-cover products such as, MERIS (Malcher et al. 2003), MODIS (Hall et al. 2002; Parajka & Blöschl 2008), SEVIRI (de Wildt et al. 2007) and ASTER (Foppa et al. 2007). However, clouds representing the similar spectral characteristics in optical satellites with snow have been the challenging point for the snow cover mapping algorithms. A combination of the theoretical model and airborne space radar (3 m to 30 cm) results in subsurface environmental exploration covering 15% of the world's surface. P band has potential to detect the ground-ice beneath the eolian surface in between 4.4 to 6.5 m. However, more studies are required in this direction towards improving the measurements of precipitation, and snow using remote sensing datasets.

Vegetation is the most important parameters by way of controlling the turbulent fluxes of sensible and latent heat between land surface and atmosphere for the distribution of net radiation energy at surface (Bosch & Hewlett 1982; Rodriguez-Iturbe et al. 1999). Heterogeneous surface makes the analysis of vegetation more complicated using satellite datasets. A better result can be obtained using satellite datasets (Curran 1980; Myneni et al. 1995; Jones & Vaughan 2010). In favor of hydro-meteorological studies, a radiometer of the size 20 cm provides information on repetitive intervals globally with suitable spatial resolution. In satellite remote sensing, the appropriation of signal characteristics with surface properties is complicated for one and can never measure in isolation the desire cause and effect in the system. In the remote sensing of vegetation from satellite-borne instruments, several challenges are there and need to be resolved for an accurate mapping of vegetation and their properties such as bidirectional effect, atmospheric effect, plant structure effect, background effect of soil, non-linear effects of scattering, effects of spatial heterogeneity, adjacency effect, non-linear mixing, topographic effects and others (Myneni et al. 1995).

◯ CHALLENGES IN SOIL MOISTURE ESTIMATION

A fundamental drawback of remotely sensed soil moisture measurement is in relating soil moisture variability at the scale of the footprint to large or small scale soil moisture variability (Piles et al. 2011b; Srivastava et al. 2013a). The resulting emissivity for soil changes from about 0.95 for dry soil to about 0.6 for wet soils at a rate of approximately 0.01% volumetric moisture content. Charpentier and Groffman in (1992) studied the effects of topography and moisture content on the variability of soil moisture within remote sensing pixels. They reported that remote sensing manifested soil moisture conditions less accurately on pixels with wide topographic variability and less precisely when the soil is dry. Mohanty et al. (2000)

have exhibited that location on the slope is crucial in establishing soil moisture variation, suggesting that a simple averaging of soil moisture values over the slope may lead to errors at different time scales. Western et al. (1999) have suggested that the presence of vegetation tends to prune the soil moisture variations caused by topography. For a majority of satellite systems, the frequency of recurrence can be a critical issue in studies involving rapidly transforming contexts such as surface soil water content. With very wide swaths it is feasible to attain twice-daily coverage with a polar orbiting satellite. A fresh alternative is the L band satellite system that will include a 1.4 GHz channel (Srivastava et al. 2014a). L band carries significant potential for estimating soil water content in regions of low and high levels of vegetation (Petropoulos et al. 2014b).

For undertaking the use of space observations, two space based systems with a 1.4 GHz (considered as L band) channel would deliver improved global soil moisture information (Kerr et al. 2001). Soil Moisture and Ocean Salinity (SMOS) and Soil Moisture Active Passive (SMAP) missions are specially launched for the soil moisture measurement and analysis at L band (Entekhabi et al. 2010; Kerr et al. 2012). A great advancement in the field of signal processing and significant components of analogue makes it more useful in the soil moisture study. Synthetic Aperture Radar has the ability to retrieve moisture at the finer scale of < 100 m. For deeper layers, the profile prediction is less accurate as the satellite can penetrate only upto few centimeters of the earth surface (Srivastava et al. 2014c; Srivastava et al. 2013d). In reality, microwaves provide instantaneous measurements of soil moisture. However, further experiments are necessary with verified measurements of surface soil moisture from remote sensing to validate and augment the findings of satellite based soil moisture retrieval (Owe et al. 2001). To represent the space-time structure of the soil moisture profile and surface fluxes partitioning, extension of the methodology over large areas is needed (Jackson & Schmugge 1989). This will require a distributed land surface model, which incorporates spatially variable atmospheric forcing and surface parameters into the model formulations (Petropoulos et al. 2014a; Zhuo et al. 2015).

⊃ CHALLENGES IN MONITORING EVAPOTRANSPIRATION ▬▬▬▬

Evaporation cannot be observed directly by Earth observation satellites. Therefore it can only be estimated from remote sensing data if combined with models that describe the land-atmosphere swaps of water and energy. Ma et al. (2009) proposed a Advanced Space-borne Thermal Emission and Reflection radiometer (ASTER) satellite and tested it over an experimental study site located on the Tibetan Plateau. The comparison with *in situ* measurements showed that the relative error of the three ASTER derived evaporation estimates was less than 10% (Kalma et al. 2008). This compares favorably with relative errors reported in the literature which are generally in the range 15-30% (Kalma et al. 2008). The discrimination between clouds and snow is still a source of uncertainty, especially over very bright surfaces. Many studies are also made on MODIS evapotranspiration product. Regarding the MODIS product, numerous validation studies have defined it to be of good quality

(Parajka & Blöschl 2008). However, cloud and atmospheric disturbances are still a problem in optical remote sensing. Gafurov & Bárdossy (2009) investigated different techniques for eliminating cloud covered pixels over the Kokcha catchment located in the north-eastern part of Afghanistan. They could remove, on average, 30% of the cloudy pixels with some accuracy.

◯ UNCERTAINTIES IN RETRIEVAL ALGORITHMS

The electromagnetic and microwave properties of the surface, physics of sensor along with connected air sea interface, understanding of geophysical process and proper checking with well spread technologies can be used for precise measurement of the variables in order to achieve desired outputs. Conversion of electromagnetic signals from satellite sensors to the measurement of climate variables is done by retrieval alogrithm. Sensor deployment is inherently non-deterministic and there is a certain degree of randomness associated with the location of a sensor in the sensor field (Zou & Chakrabarty 2004) which leads to the uncertainties in retrieval of climate trends and variables. The main reason for the apparent spectrum of change has been cited as unresolved drifts in the sensitivity of the sensor. Radiometric stability and sensitivity during the operation is slowly lost by sensors, therefore good calibration is imperative. Some satellite sensors are unable to be recalibrated after launch due to the lack of accurate on-board or on-orbit calibrations. Methods have been developed to calibrate these types of sensors but may still contain uncertainties (Xiong et al. 2003). Biases generated by instrument drifts causes errors in satellite data. When satellite passes through at the local equator there is a slow change in the crossing time, adding a spurious effect to detected trends (Mears & Wentz 2005). Such biases must be consigned by implementing a diurnal correction procedure to the data or by determining the precise orbit position of the satellites (Yang et al. 2013). A generation of long-term records can be performed by combining observations from different satellites resulting in the increase of uncertainties. If there is problem in the output of merged data from different systems during development and calibration then there would be a large increase in uncertainties. Inter-instrument calibrations can be performed to find out the relative bias and can be used to reduce such types of problems (Ohring et al. 2005).

The magnitude of detected trends is affected by uncertainties along with change in direction; for example, positive and negative trend in temperature of troposphere derived from the same satellite data by using different retrieval algorithms (Mears & Wentz 2005). An important origin of uncertainty is because of the common inputs in retrieval algorithms (Srivastava et al. 2014c). Researchers come across many options for these common inputs for obtaining good results. Evaluation of the effects of inputs on the uncertainties and evaluation of the quality of common inputs used in the retrieval algorithm need to be studied in the recent scenario. Conjointly improving retrieval algorithms and instruments, advance-quality validation data will be helpful to tune algorithms and point out the level of uncertainties in satellite data sets (Petropoulos et al. 2014b; Petropoulos et al. 2015). For achieving some of the above mentioned objectives, global reference networks for validating data products and calibrating satellite data are urgently needed. Additionally, the spatial and temporal mismatch between validation data sets and satellite observations must be noted and accounted in this process (Srivastava et al. 2015). Rigorous reanalysis advised by a better knowledge of errors in algorithms and instruments should be conducted regularly to remove errors in longterm remotely sensed data. Existing satellites and their predecessors are milestones in the producing of high quality geospatial datasets. For example, a temporally consistent global product of the Leaf Area Index has been developed using a well calibrated algorithm (Liu et al. 2012).

○ SHORT SPANS OF SATELLITE DATA AND SPATIAL RESOLUTIONS ═══

Researchers who use short time series of satellite data might have trouble in separating the long-term decadal trends and inter annual variability. Long-term continuity, consistency and homogeneity should be present in the satellite observations for accurately assessing the trends (Trenberth et al. 2013). More reliable trends can be understand by studying the length of time series of satellite datasets (Srivastava et al. 2013a). An investigation of the lengths of variables constructed from satellite observations shows that some time series are already longer than 30 years (Yang et al. 2013). The detection of robust long-term trends in some climate variables has been limited by the associated uncertainties and shortness of satellite time series. The possibility of more time series with adequate length will depend on our ability to maintain the continuity of existing satellite missions. The progress made in both instrumentation and retrieval algorithms, followed by the aggregation of satellite records, can be an antidote for this problem. A combination of the strengths of passive and active remote sensing reflects better insights into the complicated geospatial data system. Inventive use of current satellite data could leads to the generation of long-term geospatial records.

Most of the satellites are designed to provide global measurements of the variables. At a time the sensors with coarse-resolution imaging are unable to capture many processes occurring at finer spatial scales (Piles et al. 2011a). For local or regional applications, the downscaling of these data with other higher resolution data derived from other sensors are needed to bring it to a higher resolution for different applications (Merlin et al. 2008). For e.g. numerous researchers have used the optical sensors for soil moisture estimation in the past, since optical sensors provide a very high spatial resolution data (Merlin et al. 2006; Merlin et al. 2008; Piles et al. 2011a; Srivastava et al. 2013a). Sensors are not able to capture many other variables at sufficiently high resolutions because of variations in the atmospheric temperature and water vapour variability associated with small-scale turbulence (Lyapustin et al. 2011; Srivastava et al. 2014b). Similarly cloud-characteristic trends cannot be accurately measured by satellite sensors at high resolution. Thus there is a necessity for new sensors with sufficiently high spatial resolution and accuracy for observing the phenomena of interest. The deployment of large numbers of sensors could be a solution depending upon the improving capabilities of individual systems. Therefore, many new technologies are needed to achieve this.

○ CONCLUSIONS

Delivery of remotely sensed data through internet is crucial, which help users to find and extract the desired data and information in a lesser amount of time. An increasing capability and predictive power of the Earth System Model will depend on ever more complicated observing systems. Projection of decades into the future is difficult however, because future progress is restrained by a fundamental physical limit. The normal form of forthcoming remote sensing satellites and the technologies can possibly solve this problem. The spatial, spectral, radiometric and temporal domains should be of higher resolution in order to meet the requirement. For various applications, it is required to improve the spatial resolution, data accuracy, timeliness and ease-of-use of remotely sensed data. It will help in providing rapid response after occurrence of natural and human-caused disasters. Further, more work in required in developing ways to fuse data from the multi-orbiting and multi-sensors installed on different satellites dedicated for water resources development and management such as Aquarius, SMOS, SMAP, AMSR E/2 and ASCAT. Many datasets are available for water resources, even if more transparency is needed to allow various projects and programs to upgrade the quantity and quality of information available to the public. Because of issues like penetration power, satellite receptivity etc., new remote sensing technologies are needed to collect data in a more efficient and cost-effective manner so that accurate, long term datasets could be available for continuous monitoring. Further, there is need of improved algorithms and software for quick and easy processing to support programs by reputed agencies.

○ REFERENCES

Bosch, J.M. and Hewlett, J. 1982. A review of catchment experiments to determine the effect of vegetation changes on water yield and evapotranspiration. Journal of Hydrology 55(1): 3-23.

Charpentier, M.A. and Groffman, P.M. 1992. Soil moisture variability within remote sensing pixels. Journal of Geophysical Research: Atmospheres (1984–2012) 97(D17): 18987-95.

Curran, P. 1980. Multispectral remote sensing of vegetation amount. Progress in Physical Geography 4(3): 315-41.

de Wildt, M.d.R., Seiz, G. and Gruen, A. 2007. Operational snow mapping using multitemporal Meteosat SEVIRI imagery. Remote Sensing of Environment 109(1): 29-41.

Ebert, E.E., Janowiak, J.E. and Kidd, C. 2007. Comparison of near-real-time precipitation estimates from satellite observations and numerical models. Bulletin of the American Meteorological Society 88(1): 47-64.

Entekhabi, D., Njoku, E.G., O'Neill, P.E., Kellogg, K.H., Crow, W.T., Edelstein, W.N., Entin, J.K., Goodman, S.D., Jackson, T.J. and Johnson, J. 2010. The soil moisture active passive (SMAP) mission. Proceedings of the IEEE 98(5): 704-16.

Foppa, N., Hauser, A., Oesch, D., Wunderle, S. and Meister, R. 2007. Validation of operational AVHRR subpixel snow retrievals over the European Alps based on ASTER data. International Journal of Remote Sensing 28(21): 4841-65.

Gafurov, A. and Bárdossy, A. 2009. Cloud removal methodology from MODIS snow cover product. Hydrology and Earth System Sciences 13(7): 1361-73.

Gupta, M. and Srivastava, P.K. 2010. Integrating GIS and remote sensing for identification of groundwater potential zones in the hilly terrain of Pavagarh, Gujarat, India. Water International 35(2): 233-45.

Hall, D.K., Riggs, G.A., Salomonson, V.V., DiGirolamo, N.E. and Bayr, K.J. 2002. MODIS snow-cover products. Remote Sensing of Environment 83(1): 181-94.

Jackson, T.J. and Schmugge, T.J. 1989. Passive microwave remote sensing system for soil moisture: Some supporting research IEEE Transactions on. Geoscience and Remote Sensing 27(2): 225-35.

Jones, H.G. and Vaughan, R.A. 2010. Remote Sensing of Vegetation. Oxford University Press.

Kalma, J.D., McVicar, T.R. and McCabe, M.F. 2008. Estimating land surface evaporation: a review of methods using remotely sensed surface temperature data. Surveys in Geophysics 29(4-5): 421-69.

Kerr, Y.H., Waldteufel, P., Richaume, P., Wigneron, J.-P., Ferrazzoli, P., Mahmoodi, A., Al Bitar, A., Cabot, F., Gruhier, C. and Juglea, S.E. 2012. The SMOS soil moisture retrieval algorithm. IEEE Transactions on Geoscience and Remote Sensing 50(5): 1384-403.

Liu, Y., Liu, R. and Chen, J.M. 2012. Retrospective retrieval of long-term consistent global leaf area index (1981-2011) from combined AVHRR and MODIS data. Journal of Geophysical Research: Biogeosciences (2005-2012) 117(G4).

Lyapustin, A., Wang, Y., Laszlo, I., Kahn, R., Korkin, S., Remer, L., Levy, R. and Reid, J. 2011. Multiangle implementation of atmospheric correction (MAIAC): 2. Aerosol algorithm. Journal of Geophysical Research: Atmospheres (1984-2012) 116(D3).

Ma, W., Ma, Y., Li, M., Hu, Z., Zhong, L., Su, Z., Ishikawa, H. and Wang, J. 2009. Estimating surface fluxes over the north Tibetan Plateau area with ASTER imagery. Hydrology and Earth System Sciences 13(1): 57-67.

Malcher, P., Floricioiu, D. and Rott, H. 2003. Snow mapping in Alpine areas using medium resolution spectrometric sensors. *In*: Geoscience and Remote Sensing Symposium, 2003. IGARSS'03. Proceedings. 2003 IEEE International, IEEE, pp. 2835-7.

Mears, C.A. and Wentz, F.J. 2005. The effect of diurnal correction on satellite-derived lower tropospheric temperature. Science 309(5740): 1548-51.

Merlin, O., Chehbouni, A., Kerr, Y. and Goodrich, D. 2006. A downscaling method for distributing surface soil moisture within a microwave pixel: Application to the Monsoon'90 data. Remote Sensing of Environment 101(3): 379-89.

Merlin, O., Walker, J.P., Chehbouni, A. and Kerr, Y. 2008. Towards deterministic downscaling of SMOS soil moisture using MODIS derived soil evaporative efficiency. Remote Sensing of Environment 112(10): 3935-46.

Mohanty, B., Famiglietti, J. and Skaggs, T. 2000. Evolution of soil moisture spatial structure in a mixed vegetation pixel during the Southern Great Plains 1997 (SGP97) Hydrology Experiment. Water Resources Research 36(12): 3675-86.

Myneni, R., Maggion, S., Iaquinta, J., Privette, J., Gobron, N., Pinty, B., Kimes, D., Verstraete, M. and Williams, D. 1995. Optical remote sensing of vegetation: modeling, caveats and algorithms. Remote Sensing of Environment 51(1): 169-88.

Ohring, G., Wielicki, B., Spencer, R., Emery, B. and Datla, R. 2005. Satellite instrument calibration for measuring global climate change: Report of a workshop. Bulletin of the American Meteorological Society 86(9): 1303-13.

Okoye, M.A. and Koeln, G.T. 2003. Remote Ssensing (Satellite) System Technologies. Environmental Monitoring I, Encyclopedia of Life Support Systems (EOLSS). Paris: EOLSS Publishers, UNESCO. Retrieved from http://www. eolss. net.

Owe, M., de Jeu, R. and Walker, J. 2001. A methodology for surface soil moisture and vegetation optical depth retrieval using the microwave polarization difference index. IEEE Transactions on Geoscience and Remote Sensing 39(8): 1643-54.

Parajka, J. and Blöschl, G. 2008. Spatio-temporal combination of MODIS images– potential for snow cover mapping. Water Resources Research 44(3): 2008, 44, doi: 10.1029/2007WR006204.

Patel, D.P., Dholakia, M.B., Naresh, N. and Srivastava, P.K. 2012. Water harvesting structure positioning by using geo-visualization concept and prioritization of mini-watersheds through morphometric analysis in the Lower Tapi Basin. Journal of the Indian Society of Remote Sensing 40(2): 299-312.

Patel, D.P., Gajjar, C.A. and Srivastava, P.K. 2013a. Prioritization of Malesari mini-watersheds through morphometric analysis: a remote sensing and GIS perspective. Environmental Earth Sciences 69(8): 2643-2656.

Patel, D.P. and Srivastava, P.K. 2013b. Flood hazards mitigation analysis using remote sensing and GIS: correspondence with Town Planning Scheme. Water Resources Management 27(7): 2353-2368.

Patel, D.P. and Srivastava, P.K. 2014. Application of geo-spatial technique for flood inundation mapping of low lying areas. pp. 113-130. *In*: Srivastava, P.K., Mukherjee, S., Gupta, M., Islam, T., (eds.). Remote Sensing Applications in Environmental Research. Springer, Switzerland.

Petropoulos, G.P., Griffiths, H.M., Ioannou-Katidis, P. and Srivastava, P.K. 2014a. Sensitivity exploration of SimSphere land surface model towards its use for operational products development from earth observation data. pp. 35-56. *In*: Srivastava, P.K., Mukherjee, S., Gupta, M., Islam, T., (eds.). Remote Sensing Applications in Environmental Research. Springer, Switzerland.

Petropoulos, G.P., Ireland, G., Srivastava, P.K. and Ioannou-Katidis, P. 2014b. An appraisal of the accuracy of operational soil moisture estimates from SMOS MIRAS using validated in situ observations acquired in a Mediterranean environment. International Journal of Remote Sensing 35(13): 5239-50.

Petropoulos, G., Ireland, G. and Srivastava, P.K. 2015. Evaluation of the soil moisture operational estimates from SMOS in Europe: results over diverse ecosystems. IEEE Sensors Journal 15(9): 5243-5251.

Piles, M., Camps, A., Vall-Llossera, M., Corbella, I., Panciera, R., Rudiger, C., Kerr, Y.H. and Walker, J. 2011a. Downscaling SMOS-derived soil moisture using MODIS visible/infrared data. IEEE Transactions on Geoscience and Remote Sensing 49(9): 3156-66.

Piles, M., Camps, A., Vall-Llossera, M., Corbella, I., Panciera, R., Rüdiger, C., Kerr, Y.H. and Walker, J. 2011b. Downscaling SMOS-derived soil moisture using MODIS visible/infrared data. IEEE Transactions on Geoscience and Remote Sensing 49(9): 3156-66.

Reynolds, C., Yitayew, M., Slack, D., Hutchinson, C., Huete, A. and Petersen, M. 2000. Estimating crop yields and production by integrating the FAO Crop Specific Water Balance model with real-time satellite data and ground-based ancillary data. International Journal of Remote Sensing 21(18): 3487-508.

Richardson, A.D., et al. 2012. Terrestrial biosphere models need better representation of vegetation phenology: results from the North American Carbon Program Site Synthesis. Global Change Biology 18: 566-84.

Rodriguez-Iturbe, I., Porporato, A., Ridolfi, L., Isham, V. and Coxi, D. 1999. Probabilistic modelling of water balance at a point: the role of climate, soil and vegetation. pp. 3789-805. *In*: Proceedings of the Royal Society of London A: Mathematical, Physical and Engineering Sciences. The Royal Society.

Srivastava, P.K., Mukherjee, S., Gupta, M. and Singh, S. 2011. Characterizing monsoonal variation on water quality index of River Mahi in India using geographical information system. Water Quality, Exposure and Health 2(3-4): 193-203.

Srivastava, P.K., Gupta, M. and Mukherjee, S. 2012a. Mapping spatial distribution of pollutants in groundwater of a tropical area of India using remote sensing and GIS. Applied Geomatics 4(1): 21-32.

Srivastava, P.K., Han, D., Gupta, M. and Mukherjee, S. 2012b. Integrated framework for monitoring groundwater pollution using a geographical information system and multivariate analysis. Hydrological Sciences Journal 57(7): 1453-72.

Srivastava, P.K., Han, D., Ramirez, M.R. and Islam, T. 2013a. Appraisal of SMOS soil moisture at a catchment scale in a temperate maritime climate. Journal of Hydrology 498: 292-304.

Srivastava, P.K., Han, D., Ramirez, M.R. and Islam, T. 2013b. Machine learning techniques for downscaling SMOS satellite soil moisture using MODIS land surface temperature for hydrological application. Water Resources Management 27(8): 3127-44.

Srivastava, P.K., Singh, S.K., Gupta, M., Thakur, J.K. and Mukherjee, S. 2013c. Modeling impact of land use change trajectories on groundwater quality using remote sensing and GIS. Environmental Engineering and Management Journal 12(12): 2343-55.

Srivastava, P.K., Han, D., Rico-Ramirez, M.A., Al-Shrafany, D. and Islam, T. 2013d. Data fusion techniques for improving soil moisture deficit using SMOS satellite and WRF-NOAH land surface model. Water Resources Management, 27(15): 5069-5087.

Srivastava, P., O'Neill, P., Cosh, M., Kurum, M., Lang, R. and Joseph, A. 2014a. Evaluation of dielectric mixing models for passive microwave soil moisture retrieval using data from ComRAD Ground-Based SMAP Simulator. IEEE Journal of Selected Topics in Applied Earth Observations and Remote Sensing, PP(99): 1-10.

Srivastava, P.K., Han, D., Rico-Ramirez, M.A., Bray, M., Islam, T., Gupta, M. and Dai, Q. 2014b. Estimation of land surface temperature from atmospherically corrected LANDSAT TM image using 6S and NCEP global reanalysis product. Environmental Earth Sciences 72(12): 5183-96.

Srivastava, P.K., Han, D., Rico-Ramirez, M.A., O'Neill, P., Islam, T. and Gupta, M. 2014c. Assessment of SMOS soil moisture retrieval parameters using tau–omega algorithms for soil moisture deficit estimation. Journal of Hydrology 519, Part A(0): 574-87.

Srivastava, P.K., Mukherjee, S., Gupta, M. and Islam, T. 2014d. Remote Sensing Applications in Environmental Research. Springer Verlag, Switzerland.

Srivastava, P.K., Han, D., Rico-Ramirez, M.A., O'Neill, P., Islam, T., Gupta, M. and Dai, Q. 2015. Performance evaluation of WRF-Noah Land surface model estimated soil moisture for hydrological application: Synergistic evaluation using SMOS retrieved soil moisture. Journal of Hydrology 529: 200-212.

Tilman, D., Cassman, K.G., Matson, P.A., Naylor, R. and Polasky, S. 2002. Agricultural sustainability and intensive production practices. Nature 418(6898): 671-7.

Trenberth, K.E., Anthes, R.A., Belward, A., Brown, O.B., Habermann, T., Karl, T.R., Running, S., Ryan, B., Tanner, M. and Wielicki, B. 2013. Challenges of a sustained climate observing system. pp. 13-50. In: Ghassem R. Asrar and James W. Hurrell (eds.). Climate Science for Serving Society. Springer, Netherlands.

Western, A.W., Grayson, R.B., Blöschl, G., Willgoose, G.R. and McMahon, T.A. 1999. Observed spatial organization of soil moisture and its relation to terrain indices. Water Resources Research 35(3): 797-810.

Xiong, X., Chiang, K., Esposito, J., Guenther, B. and Barnes, W. 2003. MODIS on-orbit calibration and characterization. Metrologia 40(1): S89.

Yadav, S.K., Singh, S.K., Gupta, M. and Srivastava, P.K. 2014. Morphometric analysis of Upper Tons basin from Northern Foreland of Peninsular India using CARTOSAT satellite and GIS. Geocarto International 29(8): 895-914.

Yaduvanshi, A., Srivastava, P.K. and Pandey, A. 2015. Integrating TRMM and MODIS satellite with socio-economic vulnerability for monitoring drought risk over a tropical region of India. Physics and Chemistry of the Earth, Parts A/B/C. DOI: 10.1016/j.pce.2015.01.006

Yang, J., Gong, P., Fu, R., Zhang, M., Chen, J., Liang, S., Xu, B., Shi, J. and Dickinson, R. 2013. The role of satellite remote sensing in climate change studies. Nature climate change 3(10): 875-83.

Zhuo, L., Han, D., Dai, Q., Islam, T. and Srivastava, P.K. 2015. Appraisal of NLDAS-2 multi-model simulated soil moistures for hydrological modelling. Water Resources Management 29: 3503-3517.

Zou, Y. and Chakrabarty, K. 2004. Uncertainty-aware and coverage-oriented deployment for sensor networks. Journal of Parallel and Distributed Computing 64(7): 788-98.

Index

About the Editors

<<<<<<<<<<<<<<<<<<<<<<<<<<<<<<<<<<<<<<<<<<<<<<<<<<<<<<<<<<<<<<<<<<<<<<<<<<<<<<<<<<<

Prashant K. Srivastava is presently a visiting scientist in Hydrological Sciences, NASA Goddard Space Flight Center, and affiliated with IESD, Banaras Hindu University as a faculty. He received his PhD degree from Department of Civil Engineering, University of Bristol, UK. He has made significant contributions in the field of Optical/IR and microwave remote sensing, satellite soil moisture and precipitation retrieval, mesoscale and hydrological modeling etc. He has published 100+ articles in peer-reviewed journals and authored many papers in conferences. He has edited many books and written several book chapters. He is presently acting as editorial board member of several international peer-reviewed scientific journals.

Prem Chandra Pandey received PhD degree (2015) from CLCR department of Geography, University of Leicester, UK, under Commonwealth Scholarship and Fellowship Plan. He received his M.Sc. degree in Environmental Sciences from Institute of Sciences, Banaras Hindu University and M.Tech in Remote Sensing from Birla Institute of Technology, Mesra Ranchi, India. He has been a recipient of many awards including Commonwealth fellowship UK, INSPIRE fellowship, MHRD and UGC fellowships from India. His research interests include forestry, agriculture and natural resource management.

Pavan Kumar received M.Sc. degree in environmental science from Banaras Hindu University, Varanasi, India and the M.Tech. in remote sensing from the Birla Institute of Technology, Mesra, India. He is currently an Assistant Professor with the Department of Remote Sensing, Banasthali Vidyapith, Jaipur, India. His current research interests include forestry mapping, crop forecasting, water resources, climate change and urban heat island.

Akhilesh Singh Raghubanshi is currently a Professor at the Institute of Environment and Sustainable Development, Banaras Hindu University, India. He has made significant contributions in the field of integrative ecology by elucidating structure, functioning and sustainable management of tropical terrestrial ecosystems. His research primarily focuses on the understanding of plant diversity and distributions, ecology of plant invasions, mechanisms of nutrient dynamics, greenhouse gas emissions, ecohydrology, physiology of tropical trees and weed flora and restoration of degraded ecosystems.

Dawei Han is currently Professor of Hydroinformatics in the Department of Civil Engineering, the University of Bristol, UK. He obtained his BEng and MSc degrees from China and PhD from the UK. His research interests include hydroinformatics, real-time flood forecasting, flood risk assessment and management, climate change, remote sensing and Geographic Information System, water resources management, natural hazards, systems, environmental engineering.